CHEMICAL PRINCIPLES
FOR ORGANIC CHEMISTRY

Robert S. Boikess
Rutgers, The State University of New Jersey

CENGAGE
Learning·

Australia • Brazil • Mexico • Singapore • United Kingdom • United States

CENGAGE
Learning®

Chemical Principles for Organic Chemistry
Robert S. Boikess

Product Director: Mary Finch

Product Manager: Maureen Rosener

Content Coordinator: Karolina Kiwak

Marketing Brand Manager: Nicole Hamm

Rights Acquisitions Specialist: Roberta Broyer

Manufacturing Planner: Judy Inouye

Art and Design Direction, Production
Management, and Composition:
PreMediaGlobal

Cover Image: 3d abstract spheres molecules
(© Jezper/Shutterstock)

For product information and technology assistance, contact us at
Cengage Learning Customer & Sales Support, 1-800-354-9706
For permission to use material from this text or product,
submit all requests online at **www.cengage.com/permissions.**
Further permissions questions can be emailed to
permissionrequest@cengage.com.

Library of Congress Control Number: 2013944595

ISBN-13: 978-1-285-45769-7

ISBN-10: 1-285-45769-2

Cengage Learning
200 First Stamford Place, 4th Floor
Stamford, CT 06902
USA

Cengage Learning is a leading provider of customized learning solutions with office locations around the globe, including Singapore, the United Kingdom, Australia, Mexico, Brazil and Japan. Locate your local office at **international.cengage.com/region.**

Cengage Learning products are represented in Canada by Nelson Education, Ltd.

For your course and learning solutions, visit **www.cengage.com.**
Purchase any of our products at your local college store or at our preferred online store **www.cengagebrain.com.**

Instructors: Please visit **login.cengage.com** and log in to access instructor-specific resources.

Printed in the United States of America
1 2 3 4 5 6 7 17 16 15 14 13

Contents

Detailed Contents

Chapter 1

Valence Bond Theory

1.1 | The Covalent Bond

Organic compounds are covalent compounds, and with only a very few exceptions, all the bonds in organic compounds are covalent bonds. The emphasis in our picture of the covalent bond is on the sharing of electron pairs between two atoms.

When two atoms share one pair of electrons, they are attached to each other by a single bond. When two atoms share two pairs of electrons, they are attached to each other by a double bond; when they share three pairs of electrons, they are attached by a triple bond. Single bonds are by far the most common type of covalent bond.

The sharing of electrons in a covalent bond by two atoms is energetically favorable under ordinary conditions on the Earth. *Energy is released when a bond forms; energy is required to break a bond.* This simple statement is a very important general principle in organic chemistry. Stated another way, *bond making is good, bond breaking is bad.*

Some basic aspects of our picture of the covalent bond are important to keep in mind.

1. The shared electron pair lies between the two bonded nuclei because the two electrons can be attracted by both nuclei and can offset the repulsion between the nuclei themselves.
2. The two electrons of an electron pair in a covalent bond have opposite spins. This observation is a consequence of the Pauli Exclusion Principle, which states that no two electrons can have the same four quantum numbers.
3. Two atoms that form a covalent bond keep approaching each other until the energy of the system reaches a minimum. If they get any closer, the energy increases because of repulsions between the nuclei. So there is an optimum distance between the two nuclei that are bonded. This distance is called the bond length.

1.2 | Lewis Structures

Perhaps the simplest and most frequently used representations of the structures of organic molecules are Lewis structures, usually modified or simplified for convenience. As we shall see, it is especially easy to simplify Lewis structures in organic chemistry because we are dealing with relatively few and simple elements.

A correctly drawn Lewis structure indicates how all the atoms of a molecule are attached to one another and accounts for all the valence electrons of all the atoms in the molecule. The valence electrons of an atom of a nonmetal are those of highest principal quantum number. A Lewis structure does not provide any information about the shape or geometry of a molecule.

We can formulate a set of rules for drawing Lewis structures and then introduce some simplifications:

1. Only valence electrons are shown in a Lewis structure. In the elements we will encounter the number of valence electrons is equal to the group number, which is the number of the element's column in the periodic table. These numbers are summarized in the following table and should be learned.

Element	Valence electrons	Element	Valence electrons
H	1	P	5
C	4	S	6
N	5	X (F, Cl, Br, I)	7
O	6		

The word "molecules" is also intended to include polyatomic ions.

2. A shared electron pair is represented by a line between the two atoms that are sharing the pair. If two pairs are being shared, they are represented by two lines, and three shared pairs are represented by three lines. These electrons are called bonding electrons.

3. Valence electrons that are not included in covalent bonds are called nonbonding electrons and are represented by dots on the atoms to which they are assigned. Like bonding electrons, nonbonding electrons are also grouped into pairs. The dots that represent pairs of nonbonding electrons are written alongside the symbol of the element with which they are associated. They are often called lone pairs.

Lewis structures of some simple organic molecules are shown in Figure 1.1.

Figure 1.1
Lewis structures of some simple organic molecules.

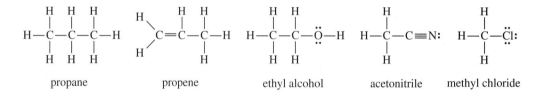

propane   propene   ethyl alcohol   acetonitrile   methyl chloride

The Octet Rule

The Lewis structures of organic molecules satisfy the **octet rule**, meaning that each atom in a molecule achieves a noble gas valence electronic configuration by electron sharing. The octet rule takes its name from the fact that all the noble gases except helium have 8 electrons in their valence shell. Because helium has only two electrons in its valence shell, hydrogen achieves a noble gas valence electronic configuration when it has a share of just two electrons.

The octet rule also means that elements in the first and second rows of the periodic table are never associated with more valence electrons than make up the corresponding noble gas valence electronic configuration. So a hydrogen atom in a molecule can *never* have more than two electrons. This restriction means that a hydrogen atom will always have one single bond and no lone pairs.

The elements of the second row of the periodic table that are found in organic molecules—carbon, nitrogen, oxygen, and fluorine—can never be associated with more than 8 valence electrons. So when atoms of these elements are in molecules, they almost always have a total of exactly four pairs of electrons, either as shared pairs in bonds or as lone pairs, never more and only very rarely fewer. We say these elements obey the octet rule.

When we consider the elements below the second row of the periodic table that are found in organic molecules—phosphorus, sulfur, chlorine, bromine, and iodine—the octet rule is usually followed, but not always. These elements have the ability to expand their octets. Under certain circumstances they can be associated with 5 or 6 pairs of electrons. These circumstances are relatively rare in organic molecules and are primarily encountered when one of these elements is bonded to oxygen.

Since almost every organic molecule that you encounter obeys the octet rule, it is a powerful tool to use in drawing Lewis structures.

Drawing Lewis Structures

Since drawing Lewis structures and simplified Lewis structures is such a fundamental skill for organic chemistry, it is useful to formulate some procedures that will help you to draw them with relative ease. We shall illustrate the application of these principles using formaldehyde, CH_2O.

Step 1: Draw the skeleton structure of the molecule by connecting all the atoms with single bonds. This initial step is usually the most important and the most challenging one.

> **a.** Some chemical information about the specific compound may be required in order to draw its skeleton structure. This information is usually experimental data that have been obtained in the laboratory and must be conveyed to us. In a problem, the information may be conveyed by just naming the compound.
>
> **b.** The octet rule limits the number of single bond connections to any atom. Hydrogen can have no more than one connection. The other elements of the second row can have no more than four connections. The elements below the second row can generally have no more than four connections unless they are bonded to oxygen or halogens.
>
> **c.** While the maximum number of connections is limited by the octet rule, each nonmetal also has a preferred number of bonds that it forms. Sometimes, as in hydrogen and carbon, that preferred number corresponds directly to the maximum permitted by the octet rule. But for the other nonmetals, the preferred number of bonds is fewer than 4. Since we will not encounter many different elements in organic chemistry, it is relatively easy to learn these preferred numbers. They correspond to 8 minus the group number of the nonmetal. So the preferred number of bonds for N and P in group VA is 8 minus 5 = 3. The preferred number of bonds for O and S in group VIA is 8 minus 6 = 2. The preferred number of bonds for the halogens in group VIIA is 8 minus 7 = 1 (see Table 1.1).
>
> **d.** In the absence of other chemical information, there are several ways in which knowing the preferred number of bonds can be helpful in writing the skeleton structure. First, we tend to avoid skeleton structures in which an atom has more than its preferred number of bonds, although they can occur if they do not violate the octet rule. Second, given a choice, it is usually better for two atoms to be equally below their preferred number of bonds, than for one atom to be further from the preferred number than the other.

For example, the skeleton structure for CH_2O is better as

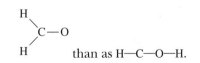

than as H—C—O—H.

In the first structure both C and O are within one of their preferred numbers of bonds (4 for C and 2 for O). In the second structure O has its preferred number, but C has two bonds fewer.

Step 2: Find the number of valence electrons in the molecule by totaling the valence electrons of each atom. Remember, in the nonmetals, the number of valence electrons is equal to the group number. For CH_2O it is 4 from the C, 2×1 from the two Hs, and 6 from the O. The total is 12.

Step 3: Count the electrons that have been used to construct the skeleton structure and subtract that number from the total number of valence electrons. If the result is zero, the structure is complete; if the result is not zero, proceed to the next step. Since the skeleton structure has 3 bonds, we have used 6 electrons out of the total of 12. So 12 minus 6 = 6 electrons are still available and must be accounted for.

Step 4: Count the number of electrons needed to complete the octets of every atom in the molecule, assuming no additional electron sharing. Since the C has 3 bonds, it needs 2 more electrons to complete its octet. Since the O has 1 bond, it needs 6 more electrons to complete its octet. A total of 8 electrons is needed to complete the octets.

Step 5: Compare the number of valence electrons available in step 3, in this case it is six, with the number needed in step 4 to complete the octets, in this case it is eight. If the number of electrons from step 3 is equal to the number from step 4, complete the Lewis structure by drawing enough dots around each atom to give it a complete octet. If the number of available electrons (from step 3) is fewer than the number of electrons needed (from step 4), as it is in this case, one additional bond must be drawn in the Lewis structure for each difference of two electrons. Such compounds will have multiple bonds or more rarely cyclic structures. Be very careful when adding bonds not to expand the octets of any atom in the structure. So formaldehyde has one double bond:

Now 6 − 2 = 4 electrons remain, and 4 electrons are needed to complete the remaining incomplete octet on O. So the structure is

Notice that the skeleton structure signaled that more bonds would be needed. Both the C and O had fewer than the preferred number of bonds.

Example 1.1

Draw Lewis structures of molecules with the following compositions:

(a) CH_5N, (b) CH_3N

Solution

(a) ***Step 1:*** Since each H can have only one bond, the C and N must be attached. It is possible to attach everything, giving C and N the preferred numbers of bonds (four and three, respectively). The skeleton structure is

Step 2: The number of valence electrons is $5H \times 1 + 1C \times 4 + 1N \times 5 = 14$.

Step 3: There are six connections (lines) in the skeleton structure, so we have used $6 \times 2 = 12$ valence electrons.

Step 4: Only the N has an incomplete octet. It has six electrons (from the three bonds) and thus needs two more.

Step 5: The difference between the number of valence electrons in step 2 and step 3 is 14 − 12 = 2 electrons available. Step 4 shows that only two electrons are needed. We can thus complete the octet of the N by adding two dots to it. The structure is

(b) **Step 1:** The skeleton structure differs from the one in part (a) by having two fewer H atoms. The best structure will be the one in which two H atoms are attached to the C and one H atom to the N. In this structure both C and N are within one of their preferred number of bonds (four and three).

$$H-\overset{\overset{\textstyle H}{|}}{C}-N\overset{\textstyle /H}{}$$

If we were instead to put all three H atoms on C and none on N, the N would be two bonds from its preferred number of three. If we were to put two H atoms on N and one on C, then the C would be two bonds from its preferred number of four. Because both C and N have fewer than their preferred number of bonds, we expect an additional bond.

Step 2: The number of valence electrons is $3H \times 1 + 1C \times 4 + 1N \times 5 = 12$ electrons.

Step 3: There are four connections in the skeleton structure, so we have used $4 \times 2 = 8$ valence electrons.

Step 4: The C has six electrons (from the three bonds) and thus needs two more. The N has four electrons (from the two bonds) and thus needs four more. A total of six more electrons are needed to complete the octets.

Step 5: The difference between the numbers in step 2 and step 3 is $12 - 8 = 4$ electrons, which is the number available to complete the octets. But step 4 shows that six electrons are actually needed to complete the octets. We must therefore add an additional bond. It must be between the C and the N, since an H cannot have more than one bond. That completes the octet of C, and the remaining two electrons are used as a lone pair to complete the octet of N.

The structure is

$$H-\overset{\overset{\textstyle H}{|}}{C}=\underset{..}{N}\overset{\textstyle /H}{}$$

Example 1.2

Draw the Lewis structure of a molecule with the composition $C_2H_2Cl_2$.

Solution

Step 1: Since the H and Cl cannot have more than one bond, there must be a bond between the two C atoms. The best structure will be one in which each C is bonded to two other atoms as well. In such a structure each C is within one of its preferred number of bonds. The other possibility in which one C has four bonds and the other has two bonds is not as good because one of the C atoms is two away from its preferred number of bonds.

> But even when we decide that the best is the skeleton structure with the C atoms attached and each having three bonds, there are still two possibilities that appear to be equally good. One has both Cl atoms bonded to the same C, and one has each Cl atom bonded to a different C.

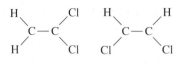

We cannot know which structure to use without additional chemical information. As is so often the case in organic chemistry, each of these skeleton structures corresponds to a different compound. Two different compounds with the same composition are called **isomers.** When the two different compounds with the same composition have different skeleton structures, they are called **constitutional isomers.**

Let us complete the Lewis structure for the first skeleton structure.

Step 2: The number of valence electrons is $2H \times 1 + 2C \times 4 + 2Cl \times 7 = $ 24 electrons.

Step 3: There are five connections (lines) in the skeleton structure, so we have used $5 \times 2 = 10$ valence electrons.

Step 4: Each C, which has three bonds, needs two more electrons to complete its octet. Each Cl, which has one bond, needs six more electrons to complete its octet. Altogether, $2 \times 2 + 2 \times 6 = 16$ electrons are required to complete the octets.

Step 5: The difference between the numbers in step 2 and step 3 indicates that $24 - 10 = 14$ electrons are available. But we need 16 electrons to complete the octets, so we need an additional bond, which must be between the two C atoms, since neither H nor Cl can have more than one bond. The additional bond completes the octet of each C atom. The remaining 12 electrons are used as lone pairs to complete the octet of each Cl.

The structure is

We could complete the Lewis structure of the second skeleton structure in the same way. It is

Structures with Charges

Many chemical species have Lewis structures that have one or more atoms with an electrical charge.

When a chemical species has a net charge, it is called an ion. The reason that a polyatomic ion has a charge is because its number of valence electrons is either greater than or fewer than the total number of valence electrons of its constituent atoms. If the ion has fewer valence electrons, then it bears a positive charge and is called a **cation.** If it has more valence electrons, it bears a negative charge and is called an **anion.**

The Lewis structures of polyatomic ions are formulated following the same steps we have used for neutral molecules. Sometimes in cations, it will not be possible to complete all the octets.

Consider the structure of the methyl cation CH_3^+.

The skeleton structure must be

H\\
      C
H/   \\H
      |
      H

Its charge of $+1$ indicates that it has one valence electron fewer than the total number of valence electrons of its atoms. The number of valence electrons from its atoms is $3H \times 1 + 1C \times 4 = 7$ electrons. So there are six valence electrons available, and we have used them all. The carbon has an incomplete octet. We complete the structure by putting the positive charge next to carbon.

$$
\begin{array}{c}
H \diagdown \overset{+}{C} \diagup H \\
| \\
H
\end{array}
$$

Many cations will not have an incomplete octet, however. Consider the structure of the methyl ammonium ion, CH_6N^+.

The skeleton structure must be

$$
\begin{array}{cc}
H & H \\
| & | \\
H-C-N-H \\
| & | \\
H & H
\end{array}
$$

The number of valence electrons from the atoms is $6H \times 1 + 1C \times 4 + 1N \times 5 = 15$. The actual number is 14, one fewer because of the $+$ charge. We have used 14 electrons for the skeleton structure. In fact, it is not possible for this species to have a 15th electron because there is no place to put it without violating the octet rule. Thus this species has 14 valence electrons and a $+$ charge.

$$
\begin{array}{cc}
H & H \\
| & | \\
H-C-\overset{+}{N}-H \\
| & | \\
H & H
\end{array}
$$

The positive charge is placed on the N, not on the C because the N has one more than its preferred number of bonds. When an atom has one more than its preferred number of bonds and a completed octet, it has a charge of $+1$. The N in this ion or the O in H_3O^+ are examples. When an atom, such as the C in this ion, has its preferred number of bonds and a completed octet, it does not have a charge.

When an atom has fewer than its preferred number of bonds and has a completed octet, it has a negative charge. If it has one bond fewer, it has a charge of -1. If it has two bonds fewer, it has a charge of -2. Notice that the C in the methyl cation has only three bonds. But it has a positive charge, rather than a negative charge because it does not have a completed octet.

Incomplete octets will not be an issue in anions. But when drawing the structures of anions, which have more valence electrons than their atoms bring, we must be careful not to violate the octet rule by expanding octets. In all but the most complicated cases, octet expansion will not be a concern.
Consider the structure of the methyl anion CH_3^-.

The skeleton structure must be

$$
\begin{array}{c}
H \diagdown C \diagup H \\
| \\
H
\end{array}
$$

Its charge of -1 indicates that it has one more valence electron than the total number of valence electrons of its atoms. The number of valence electrons from its atoms is $3H \times 1 + 1C \times 4 = 7$. So there are $7 + 1 = 8$ valence electrons, and we have used only 6. The carbon has an incomplete octet and needs two more electrons. We assign the two electrons to carbon as a lone pair. The structure is

Examination of the structures of these ions reveals a feature that is found in the structure of every ion. At least one atom does not have its preferred number of bonds. In the ions of carbon, the C has only three bonds, when four is preferred. In the ammonium ion the N has four bonds, when three is preferred.

The structural formulas of neutral molecules may also have charges. But in order to preserve neutrality, there must be equal numbers of positive and negative charges. These charges are associated with atoms that do not have their preferred number of bonds. It is not always possible to assign the preferred number of bonds to every atom in a structure. Its composition or experimental evidence about its properties may require a structure in which some atoms do not have the preferred number of bonds.

Example 1.3

Draw the Lewis structure of nitromethane, CH_3NO_2.

Solution

Step 1: There appear to be several possible skeleton structures. Chemical information and experience allow us to select the correct one. It is the one that has the structure of a nitro group. A nitro group has two O atoms attached separately to an N atom, O—N—O, and the N atom is attached to the rest of the molecule.

The skeleton structure is

Notice that in this skeleton structure, only the O atoms do not have the preferred number of bonds.

Step 2: The number of valence electrons is $3H \times 1 + 1C \times 4 + 1N \times 5 + 2O \times 6 = 24$ electrons.

Step 3: There are six connections in the skeleton structure, so we have used $6 \times 2 = 12$ valence electrons.

Step 4: The octet of C is complete; the N requires 2 electrons; and each O requires 6 electrons for a complete octet, a total of 14 electrons. But only $24 - 12 = 12$ electrons are available. So an additional pair of electrons needs to be shared. It can be placed only between the N and an O. The remaining 10 electrons are then sufficient to complete the octets of the two O atoms.

The structure is

In this structure the N and one of the O atoms do not have the preferred number of bonds, although they all have completed octets. This drawing places the double bond between the N atom and the upper O atom. The double bond could just as well have been placed between the N atom and the lower O atom. We shall see the importance of this observation when we discuss resonance.

Every atom in a molecule or ion can be assigned what is called a **formal charge.** A formal charge does not indicate the charge distribution in a molecule. It is simply a device that helps us with electron bookkeeping and is based on how many valence electrons the atom has, how many pairs of electrons the atom shares, and how many lone pairs it has. Most atoms in most molecules or ions have a formal charge of zero and we ignore them. The sum of all the formal charges in a molecule must be zero. The sum of all the formal charges in an ion must be the charge of the ion. We can calculate the formal charge on an atom in a molecule or ion using a fairly complicated general formula. But it is not necessary for us to use this formula because we can relate the formal charge of an atom to the number of bonds it has in the Lewis structure. If it has the preferred number of bonds, it has a formal charge of zero. If it does not have the preferred number of bonds, the atom will have a formal charge.

You can see the relationship between the number of bonds and the formal charge by looking at the previous examples. As we have seen, if an atom has one more bond than the preferred number and has a *completed octet*, it has a formal charge of +1. Examples are N with four bonds in the ammonium ion or nitromethane and O with three bonds in the hydronium ion.

If an atom has one bond fewer than the preferred number and has a *completed octet*, it has a formal charge of −1. Examples are the O in nitromethane and the C in the methyl anion. If an atom has two bonds fewer than the preferred number and has a completed octet, it has a formal charge of −2.

The Lewis structure of nitromethane in the previous example should be written showing the formal charges.

Because there are so few different elements that make up the organic compounds we will study in the first year, we can easily list the bonding arithmetic for each element with a *completed octet*, as shown in Table 1.1.

Although formal charges do not indicate actual charge distribution in a molecule or polyatomic ion, they are very useful as a guide to writing correct skeleton structures and for evaluating the relative stability of closely related species.

Formal charges signal instability. Therefore, we avoid them if we can. So when choosing a skeleton structure, absent other information, we choose the one that leads to the least formal charges. It will tend to be the one in which each atom comes as close as possible to its preferred number of bonds. It is for this reason that when we draw skeleton structures, it is usually better for two atoms to be equally below their preferred number of bonds, than for one atom to be further from the preferred number than the other.

When two different compounds have the same composition, the one with the least formal charge will be the more stable. For example, there are two compounds with the composition CHN. One of them is hydrogen cyanide, whose Lewis structure H—C≡N: has no formal charges. The other one is hydrogen isocyanide, whose Lewis structure is H—N⁺≡C:⁻

Table 1.1	Bonding Arithmetic for Completed Octets				
Element	**Preferred number of bonds**	**Maximum number of bonds**	**Formal charge**	**Fewer number of bonds**	**Formal charge**
C	4	4		3	−1
H	1	1		No	
N and P*	3	4	+1	2	−1
				1	−2
O and S*	2	3	+1	1	−1
F	1	1		No	
Cl, Br, I*	1	2	+1	No	

* In their lowest oxidation states

It has formal charges of +1 and −1. It is less stable than hydrogen cyanide.

In addition to information given about a specific compound, general chemical knowledge can also be very helpful when drawing Lewis structures.

Example 1.4

Draw the Lewis structure of formic acid, CH_2O_2.

Solution

Step 1: A number of reasonable skeleton structures are possible, but only one does not have an O—O bond. As a rule, we do not connect oxygen atoms to one another in a skeleton structure unless there is specific information that such bonds exist in the compound. The reason is that O—O bonds are relatively weak. Very often we will need chemical information of this kind to draw the correct skeleton structure. In addition, the identification of the molecule as formic acid tells us that there should be an O—H bond because oxyacids have O—H bonds.

The skeleton structure is

$$\begin{array}{c} H\diagdown \quad \diagup O-H \\ C \\ | \\ O \end{array}$$

Step 2: The number of valence electrons is 2H × 1 + 1C × 4 + 2O × 6 = 18 electrons.

Step 3: There are four connections in the skeleton structure, so we have used 4 × 2 = 8 valence electrons.

Step 4: The C has six electrons (from the three bonds) and thus needs two more. One O has four electrons (from the two bonds) and thus needs four more. The other O has two electrons (from the one bond) and thus needs six more. A total of 12 more electrons are needed to complete the octets.

Step 5: The difference between the numbers in step 2 and step 3 is 18 − 8 = 10 electrons. But 12 electrons are needed to complete the octets. We must therefore add an additional bond between the C and the O, which completes the octet, of C and does not introduce formal changes. The octet of each O is completed with two lone pairs. The structure is

$$\begin{array}{c} H\diagdown \quad \diagup \overset{\cdot\cdot}{\underset{\cdot\cdot}{O}}-H \\ C \\ \| \\ :\overset{}{O}: \end{array}$$

It has no formal charges.

Example 1.5

Draw the Lewis structure of formamide, CH_3NO.

Solution

Step 1: While there are a number of plausible skeleton structures, the identification of the compound as an amide indicates to a student of organic chemistry that both the N and the O are bonded to the C and that there is no O—H bond. The skeleton structure is

$$\begin{array}{c} O \qquad\quad H \\ | \qquad \diagup \\ H-C-N \\ \qquad\quad \diagdown \\ \qquad\qquad H \end{array}$$

Step 2: The number of valence electrons is 3H × 1 + 1C × 4 + 1N × 5 + 1O × 6 = 18 electrons.

Step 3: There are five connections in the skeleton structure, so we have used $5 \times 2 = 10$ valence electrons.

Step 4: The C has six electrons (from the three bonds) and thus needs two more. The O has two electrons (from the one bond) and thus needs six more. The N has six electrons (from the three bonds) and thus needs two more. A total of 10 more electrons are needed to complete the octets.

Step 5: The difference between the numbers in step 2 and step 3 is $18 - 10 = 8$ electrons. But 10 electrons are needed to complete the octets. We must therefore add an additional bond. It is best between the C and the O, since it gives each atom its preferred number of bonds and the N already has three bonds, its preferred number. The octet of the O is completed with two lone pairs, and the octet of the N is completed with one lone pair. The structure is

1.3 | Resonance

In drawing Lewis structures, there is often more than one place to put an extra bond to complete the octets. For example, in nitromethane (Example 1.3) the structure could have been drawn in two equivalent ways.

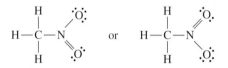

Both equivalent structures are consistent with the octet rule and our chemical knowledge. But neither is correct, if the criterion is an accurate representation of the actual molecule.

A similar choice arose when we considered the structure of formamide (Example 1.5). We put the extra bond between the C and the O, which gave each atom its preferred number of bonds and avoided formal charges. But there is another place where we could have put the extra bond. It is between the C and the N. This location is clearly not as good as the one we chose because it gives the N four bonds and the O one bond and results in formal charges. The structure would be

Here too, neither structure is correct, even though one is clearly better.

In a compound where there is more than one reasonable location for a multiple bond, no *single* Lewis structure can describe the molecule. More than one structure must be drawn for an accurate description. Nevertheless, we will draw single Lewis structures for such molecules because the Lewis structures still convey a great deal of useful information.

In order to go from these Lewis structures to an accurate description of the molecule, we need to add some principles to the ordinary Lewis structure method. These principles are summed up in the theory of resonance, which gives a method for accurately representing molecules that cannot be represented by a single Lewis structure:

1. If two or more Lewis structures that differ only in the distribution or number of multiple bonds can arise from the same skeleton structure, **resonance** is said to exist.

2. Such Lewis structures (those for nitromethane or for formamide shown earlier are examples) are called **contributing structures.** The molecule is said to be a **resonance hybrid** of its contributing structures.

3. A contributing structure is imaginary; it does not represent a real molecule. It only serves to help us draw the structure of a molecule that is a resonance hybrid.

4. The structure of a molecule (or ion) that is a resonance hybrid is described as a *blend* of its contributing structures. The structure of a resonance hybrid is represented by writing its contributing structures separated by a double-headed arrow ⟷. This symbol does *not* mean that the molecule flips back and forth between its contributing structures. Rather it means that the real structure of the molecule is a blend of the contributing structures, which are imaginary.

Thus we represent the structure of nitromethane, which is a resonance hybrid, by drawing the two equivalent contributing structures.

The real nitromethane is a blend of these two imaginary contributing structures. Since the two structures are equivalent nitromethane must be an equal blend. An equal blend of these two contributing structures describes a molecule in which each of the bonds between N and O is the same and each of these bonds is exactly intermediate between a single bond and a double bond. Put another way, the extra shared electron pair is **delocalized**. It is associated with three atoms of the molecule (the N and the two Os). An equal blend also describes a molecule in which the two O atoms are identical, each having one half of a formal negative charge.

Using the same approach we can represent the structure of formamide, which is a resonance hybrid, by drawing the two contributing structures.

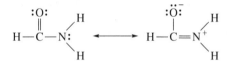

The real formamide is a blend of these two imaginary contributing structures. Since the first structure, the one without the formal charges, is better than the second one, we expect the blend to be unequal. The first structure will make a greater contribution to the description of the real structure.

The blend describes a molecule in which the bond between the C and the O is less than a full-fledged double bond, but not as much less as the N=O bond in nitromethane. The blend also indicates that the bond between the C and N is not an ordinary single bond, but that it has some double bond character. And finally there is some partial formal charge on the N and O.

In one sense a molecule that is a resonance hybrid is like a mule, which is a blend of a horse and a donkey, but does not flip back and forth between them. But this analogy is not perfect because unlike contributing structures, which are imaginary, both a horse and a donkey are real. A better analogy is to describe a rhinoceros as a blend of a unicorn and a dragon, both of which are imaginary.

One important consequence of resonance is that a resonance hybrid is more stable than we expect on the basis of any one of its contributing structures. The extra stability is called its **resonance energy**. It is most important when the contributing structures are equivalent or close to equivalent. Resonance energy is an important property of a chemical species. Resonance can often help us understand physical and chemical behavior. Conversely, the behavior of a molecule or ion can indicate that it may have more than one contributing structure.

Many molecules have some contributing structures that are much less important than others. These minor contributing structures make only small contributions to the description of a molecule, and we tend not to draw them. Sometimes, however, even a minor contributing structure can help explain an important property, and then we will reference the contributing structure in our explanation.

In order to assess the relative importance of a contributing structure, we pretend it is real and evaluate its stability, using the criteria that are used to evaluate the relative stability of chemical structures. These criteria are

1. *The number of bonds.* It is usually the most important criterion. Other things being essentially equal, the structure with the greater number of bonds will be the more stable.
2. *Charge.* Structures with charges are less stable than structures without charges. The more charge the less stable.
3. *Distribution of charge.* Structures with negative charge on more electronegative atoms are better. Placing negative charge on less electronegative atoms makes a structure less stable.
4. *Incomplete octets.* A structure in which one or more atoms have incomplete octets is less stable than the one in which all the octets are complete.

We can illustrate these ideas with formaldehyde, CH_2O, whose skeleton structure we have already discussed. By far its most important contributing structure is

But we will sometimes consider a much less important contributing structure

It is much less important because it has one bond fewer, two charges, and an incomplete octet on the C. But it does not look completely unreasonable based on our general chemical experience. It makes a minor contribution that helps explain the observation that many anions or even neutral species with lone pairs react with formaldehyde and attach to the C. You can think of still another contributing structure that we do not consider at all. It looks very unreasonable because it gives the very electronegative O the positive charge and the incomplete octet.

We can provide some guidance about when we generally do not need to consider a contributing structure:

1. Contributing structures with two formal charges in which the more electronegative element bears the positive charge and the less electronegative one bears the negative charge are rarely considered.
2. Contributing structures with incomplete octets on very electronegative elements such as N, O, F, and Cl are rarely considered.
3. Contributing structures with formal charges greater than +1 or −1 are rarely considered.
4. Contributing structures in which there is an expanded octet on H, C, N, O, or F are never considered.

Example 1.6

Draw the three most important contributing structures of C_3H_4O and identify the most important one.

Solution

Step 1: There are a number of skeleton structures that are possible. Let us consider only one of them in which every atom is within one of its preferred number of bonds. It is in this one that it will most likely be possible to write at least one contributing structure without formal charges. This skeleton structure is

$$H-C-C-C-O$$
$$\quad\ \ |\quad\ |\quad\ |$$
$$\quad\ \ H\quad H\quad H$$

Step 2: The number of valence electrons is 4H × 1 + 3C × 4 + 1O × 6 = 22 electrons.

Step 3: There are seven connections in the skeleton structure, so we have used 7 × 2 = 14 valence electrons.

Step 4: Each of the three Cs has six electrons (from the three bonds) and thus needs two more. The O has two electrons (from the one bond) and thus needs six more. A total of 3 × 2 + 6 = 12 more electrons are needed to complete the octets.

Step 5: The difference between the numbers in step 2 and step 3 is 22 – 14 = 8 electrons. But 12 electrons are needed to complete the octets. We therefore must add two additional bonds. We can add one between the C and the O and one between two Cs. Then we can complete the octet of O with the two remaining lone pairs. The structure is

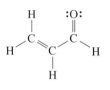

But suppose that we place the shared pair between the C and the O as a lone pair on the O. The resulting structure, in which that C has an incomplete octet and a positive charge and the O has only one bond and a negative charge, is not as important as the previous structure. But it is still plausible because of the electronegativity of the O.

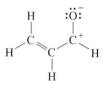

Still another contributing structure could have the additional bond between the two middle C atoms. In that case the end C would have the positive charge and incomplete octet and the O would have the negative charge.

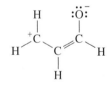

This contributing structure is also less important than the first one.

Although they are both less important, the last two contributing structures help us understand some of the properties of this molecule, which is called acrolein. The total picture they give is one in which two electron pairs are delocalized over four atoms (three Cs and an O). We say such a molecule is *conjugated*. This picture correctly predicts the observed greater stability of acrolein, compared to other similar molecules. It correctly predicts that acrolein, like formaldehyde, reacts at the C next to the O with anions or even neutral species with lone pairs. Most strikingly, it correctly predicts that anions or even neutral species with lone pairs can react with acrolein and attach to the end C as well.

Being able to draw the reasonable contributing structures of a molecule is a very important skill for organic chemistry. The method we have used so far will work, but it is slow and tedious. We need a more efficient method.

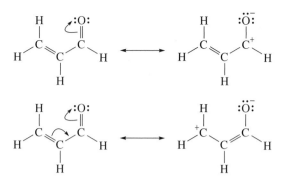

Figure 1.2

Using curved arrows to indicate electron movement.

The approach is to start with one of the contributing structures, usually an important one, that you may have to draw or that you may be given. Then without *changing the skeleton structure in any way,* determine the possible alternate locations for the extra shared pairs and lone pairs of electrons by moving them around. The extra shared pairs are represented by the second line of a double bond or the second and third lines of a triple bond.

Because the skeleton structures cannot change in any way, you cannot move an electron pair toward or away from any atom that has four single bonds. Because of the prohibition on octet expansion of H, C, N, O, or F, if you move an electron pair to one of these atoms, you must at the same time move one away. The limitations that eliminate unreasonable contributing structures from consideration are always understood.

What we need is a simple way to do our electron bookkeeping as well as indicate electron movement. The device that is used is a curved arrow as shown in Figure 1.2.

Such arrows are used in organic chemistry to account for real or imaginary (as in this case) movement of electrons. The tail of the arrow is on an electron pair in a given location in a structure. The head of the arrow points to its new location. The double head of the arrow indicates that we are referring to a pair of electrons. The result of an electron pair movement, shown by a curved arrow, is to increase by 1 the charge on the atom at the location of the tail and to either decrease by 1 the charge on the atom at the location of the head or create a multiple bond to that atom. These arrows are used not only to generate contributing structures, as shown here, but also to explain how reactions take place, what are called reaction mechanisms.

Here the movement of the electron pairs is imaginary because a contributing structure is imaginary, but using these arrows helps us write the various important contributing structures of a species. Being able to draw and use these arrows for identifying and drawing contributing structures is an essential skill.

Example 1.7

Draw the two other most important contributing structures of the cation shown below and indicate which of the three contributing structures is the most important and why.

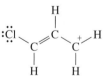

Solution

Since the most important factor in determining the importance of a contributing structure is its number of bonds, we will not consider structures with bonds fewer than this one. Movement of the extra shared pair of electrons between the two C atoms toward the positive charge gives another important contributing structure.

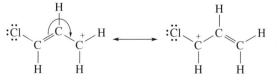

Note that the positive charge is no longer on the C on the right because this C is now sharing a pair of electrons that it did not have in the other structure and it has a complete octet with four bonds. The positive charge is now on the C from which the shared pair moved. Its loss of the shared pair results in a positive charge and an incomplete octet with three bonds.

We can move another electron pair, this time a lone pair on the Cl, toward the positive charge again.

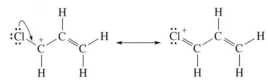

The resulting structure is the most important contributing structure because it has one more bond, the one between Cl and C, than either of the other two. This extra bond more than makes up for the two features of the structure that are bad. There is a positive charge on the electronegative Cl, and the Cl has one more than its preferred number of bonds.

1.4 | Orbitals

Lewis structures give us very important information about molecular structure. They tell us what is attached to what. They tell us something about charge distribution. In conjunction with the theory of resonance, they tell us about relative stability. But they tell us virtually nothing about the shape of molecules, or about how electrons are shared, nor do they provide information about the actual spatial distribution of the valence electrons around the atoms.

These limitations of Lewis structures are the result of the way electrons are described, or rather not described in Lewis structures. Electrons are represented as dots or dashes. But the position of an electron in space cannot be defined precisely, as dots or dashes imply, because of the wave nature of the electron. The best that can be done is to define the probability for an electron being at a given location. A simple graphical device has been created for describing this probability. It is called an **orbital.** An orbital is a picture of the probability of finding an electron at different points in space.

One simple and useful way of presenting this picture is to make a three-dimensional plot of all the points in space where the probability of finding an electron has an arbitrarily assigned constant value. Such a plot is a continuous closed surface, called a surface of constant probability, which encloses a region of space. We would like this region of space to be the region in which the electron is likely to be most of the time. So we choose a value for the probability that defines such a region.

An orbital is a convenient picture we draw to describe the electron. It is not a physical reality. If we could look at an atom or a molecule we would not see orbitals. But orbitals can be used to describe the electrons in a chemical species and to predict, usually with substantial success, the physical and chemical properties of that species.

It all starts with atoms. As we know, we describe electrons in atoms as being in atomic orbitals. These atomic orbitals come in various shapes, sizes, and energies. Since the compounds we will study consist primarily of second-row nonmetals, there are only two kinds of orbitals for us to consider: **s** orbitals and **p** orbitals. All their valence electrons are in these two kinds of orbitals. Their shapes are shown in Figure 1.3.

The energy and size of the orbitals that describe the valence electrons of an element depend on the location of the element in the periodic table. Hydrogen in the first row has the smallest and lowest energy orbital. The 1, called the principal quantum number, in its 1s designation conveys this information.

The orbitals that describe the valence electrons in the second-row elements are the 2s and the three 2p orbitals, $2p_x$, $2p_y$, and $2p_z$. They are larger and of higher energy than the 1s orbital of hydrogen, as conveyed by their principal quantum number 2. The 2p orbitals are also slightly higher in energy than the 2s orbital. The shapes of the orbitals also indicate that an electron in a p orbital is likely to be further from the nucleus than one in an s orbital. The subscripts x, y, and z indicate that the three 2p orbitals point in these three directions and are therefore perpendicular to each other, as shown in Figure 1.3.

We can use these ideas about atomic orbitals along with Lewis structures to develop a more useful picture or model of molecular structure. Covalent bonds can be understood better if the description of the electrons of these bonds is based on the concept of orbitals. The basic approach is straightforward. A covalent bond between two atoms is the result of the overlap or combination of a valence shell orbital from each. So we should be able to describe various characteristics of a given bond by considering what we know about the shape, size, and energy of the orbitals that overlap to form this bond.

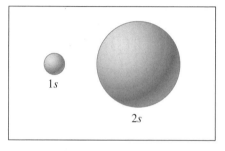

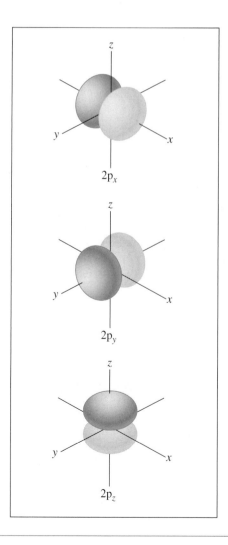

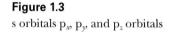

Figure 1.3
s orbitals p_x, p_y, and p_z orbitals

Unfortunately, when we examine molecules experimentally, we quickly find that this concept of overlapping atomic orbitals gives an oversimplified and inaccurate picture of molecular structure. The need for modifying the simplified model is evident even in a molecule as simple as methane, CH_4. The Lewis structure of methane is

We know that each H has a 1s orbital in its valence shell and that the C has four orbitals in its valence shell, a 2s and three 2p valence orbitals, suggesting that there should be two types of C—H bonds. In three of these bonds the 1s orbital of a hydrogen should overlap with a 2p orbital of carbon. The fourth bond should be formed by the overlap of the 1s orbital of a hydrogen with the 2s orbital of carbon. Since two different kinds of carbon orbitals with different shapes, sizes, and energies are used to form the four bonds of methane, methane should have two different kinds of bonds. There should be three of one kind that use 2p orbitals and one of another kind that uses the 2s orbital.

Every experimental result obtained for the structure of methane indicates that all four bonds in methane are identical. The simple model of overlap of s and p atomic orbitals is therefore wrong. But we can make some relatively simple modifications of this model that explain the experimental results.

1.5 | Hybridization

The hybridization model accounts for the experimental evidence about molecular structure much better than the simple overlap model. It does so by modifying the description of atomic orbitals when they overlap to form bonds. We do not describe the electrons by using s and p orbitals, as we do in atoms. Instead, we use **hybrid orbitals**, which are combination orbitals formed by the mixing of simple atomic orbitals. This modified description can be used successfully to predict molecular geometry as well as a number of bond properties.

The mixing or combining of atomic orbitals to form hybrid orbitals must obey certain rules:

1. The number of hybrid orbitals equals the number of atomic orbitals that are combined to make them. The number of orbitals associated with any chemical species remains constant no matter how we choose to combine or blend them.
2. The total energy of the hybrid orbitals is equal to the total energy of the atomic orbitals that are combined to make them. The energy of a chemical species does not change just because we choose to describe it differently.

Although it may appear that hybrid orbitals are nothing more than a different way of describing electrons, there is an underlying physical reality that we are actually describing. When electrons are shared between two atoms, in other words, when they form bonds, their distribution in space changes and the shapes of the orbitals we use to describe them also change. The new distributions and the new shapes are ones that are more suited for forming bonds.

A bond is the result of the overlap of two orbitals, one from each atom of the bond. The better the overlap, the stronger the bond. Pure atomic orbitals are quite symmetrical. As a result, when two atomic orbitals on two different atoms come together, there is relatively little overlap before their nuclei approach close enough to begin repelling each other. This relatively poor overlap is shown in Figure 1.4(a) for two s orbitals overlapping or two p orbitals overlapping. Hybrid orbitals are less symmetrical than unhybridized ones. They "point," meaning they have high electron density in one direction. When two hybrid

Figure 1.4
(a) s orbital overlap and p orbital overlap
(b) hybrid orbital overlap

orbitals point at each other, more effective overlap is the result, as shown in Figure 1.4(b).

Four Hybrid Orbitals

Let us start by using the hybridization model to describe methane, CH_4. All four bonds in methane are identical and are bonds between the C atom and one of the H atoms. We can account for the four identical bonds in CH_4 by proposing that the four orbitals in the valence shell of the C mix to form four new equivalent hybrid orbitals. The four hybrid orbitals of the C are formed by the mixing of one 2s orbital and three 2p orbitals. Therefore each one of the four hybrid orbitals is composed of one part of s orbital and three parts of p orbital.

The hybrid orbital is named by indicating the contribution of each atomic orbital with a superscript, omitting the superscript 1. The four hybrid orbitals of the C of methane are called $2sp^3$ orbitals. The superscripts are 1 (omitted) and 3. They indicate that each of these four hybrid orbitals contains one part of s orbital and three parts of p orbital. The 2 is the principal quantum number of the valence shell of the C. The principal quantum number is usually omitted from the hybrid orbital designation. We will generally follow this practice.

It can be shown mathematically that such orbitals are quite directional with most of their electron density pointed in one direction. It can also be shown that these four orbitals point toward the four corners of a geometric solid that has four faces that are equilateral triangles of the same size. This solid is called a regular tetrahedron. The C lies at the center of the tetrahedron, and lines from the C to the corners of the tetrahedron form angles of 109.5°. Figure 1.5 shows these four hybrid orbitals.

The description of CH_4 using fours p^3 hybrid orbitals is quite consistent with the experimental evidence. Not only are the four bonds equivalent, but the observed geometry is the same as that predicted by the hybridization. We draw lines between the C atom and each of the four H atoms in methane, just as if we were drawing its Lewis structure. Any two of those intersecting lines form an angle called a *bond angle*. In methane all the bond angles are 109.5° 28'. Thus we can say that methane has ideal tetrahedral geometry, as shown in Figure 1.6.

Three Hybrid Orbitals

The description of methane uses four hybrid orbitals because the C is bonded to four other atoms. Suppose that we have a compound such as C_2H_4, in which each C is bonded to only three other atoms.

Now only three hybrid orbitals are necessary for the C to bond to the three other atoms. We make these three hybrid orbitals from one 2s orbital and two 2p orbitals. Again, we see that the number of orbitals used to make the hybrid orbitals is equal to the number of hybrid orbitals made. These orbitals are named using exactly the same method we used for the sp^3 orbitals. Each of these three orbitals has one part of 2s orbital and two parts of 2p orbital. So they are called sp^2 orbitals.

Just as for sp^3, it can be shown mathematically that such orbitals are quite directional with most of their electron density pointed in one direction. It can also be shown that these three orbitals point toward the three corners of an equilateral triangle and that the C atom is in the center of the triangle. As shown in Figure 1.7, this geometry means that the three atoms bonded to C by the three sp^2 hybrid orbitals as well as the C itself all lie in the same plane. This planar arrangement means that the bond angles are 120°. This kind of ideal geometry is called trigonal planar; but as we will discuss, the geometry of C_2H_4 is close to, but not exactly, the ideal.

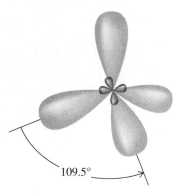

Figure 1.5
$2sp^3$ hybrid orbitals/regular tetrahedron

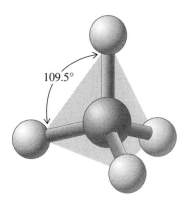

Tetrahedral

Figure 1.6
Tetrahedral geometry of methane

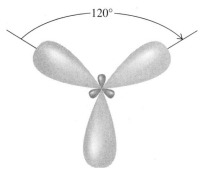

Figure 1.7
sp^2 hybrid orbitals/trigonal planar

Two Hybrid Orbitals

Suppose that we have a C bonded to two other atoms in a compound such as C_2H_2.

$$H—C≡C—H$$

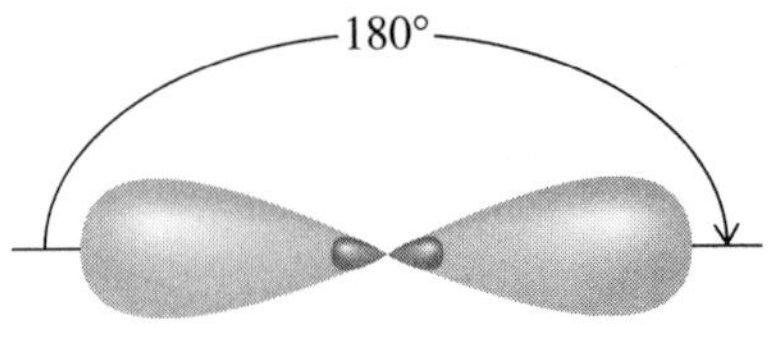

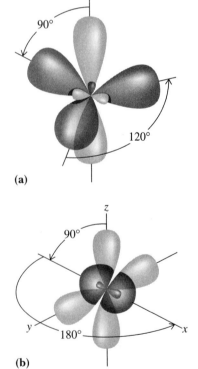

Figure 1.8

sp hybrid orbitals/collinear

Now only two hybrid orbitals are necessary in order to bond each C to the other two atoms. We make these two hybrid orbitals from the 2s orbital and one 2p orbital. These two orbitals, each of which is one part 2s orbital and one part 2p orbital, are called sp orbitals.

It can be shown mathematically that such orbitals are quite directional with most of their electron density pointed in one direction. It can also be shown that these two orbitals lie on the same line and point in exactly opposite directions, thus forming an angle of 180°, as shown in Figure 1.8.

Determining hybridization by counting atoms bonded to a C generally works well because bonded C does not have lone pairs. But elements such as N and O generally do have one or two lone pairs when they are part of molecules. To determine the hybridization of an atom with one or two lone pairs, a lone pair is counted as if it were another bonded atom. For example, the N in CH_5N, shown in Example 1.1(a), is bonded to three atoms (one C and two Hs) and has a lone pair, so it is said to be sp^3 hybridized. The N in CH_3N, shown in Example 1.1(b), is bonded to two atoms and has a lone pair. It is sp^2 hybridized.

Sigma Overlap and Pi Overlap

When two hybrid orbitals each on a different atom point at each other, they are also pointing at the nucleus of the other atom. This arrangement is shown in Figure 1.4(b) and results in the most effective overlap between orbitals on two different atoms. The overlap is along the internuclear line, an imaginary line drawn between the nucleus of each of the atoms of the bond. The most effective overlap results in the best and strongest bond between the two atoms. This type of overlap is called *sigma (σ)* overlap, and the bond that results is called a *sigma bond*. There can be only one sigma bond between two atoms.

But we have already seen that two pairs of electrons in a double bond or even three pairs of electrons in a triple bond can be shared between two atoms. Since only one sigma bond is possible between two atoms, because the two electrons completely occupy the space between the two nuclei, the second and third shared electron pairs must form another kind of bond, using other orbitals. If we do the atomic orbital bookkeeping for the formation of sp^2 and sp hybrid orbitals, these other orbitals are easily identifiable.

We combine three atomic orbitals to make the three sp^2 hybrid orbitals needed to bond three atoms. Left over is a single p orbital. It remains unhybridized and is perpendicular to the plane of the three sp^2 hybrid orbitals, as shown in Figure 1.9(a). Or, we combine two atomic orbitals to make the two sp hybrid orbitals needed to bond two atoms. Left over are two p orbitals. They remain unhybridized and are perpendicular to the line of the two sp orbitals as well as to each other, as shown in Figure 1.9(b).

In our picture of multiple bonds, both double bonds and triple bonds, it is the unhybridized p orbitals that overlap to form the second bond of a double bond or the second and third bonds of a triple bond.

The relationships between the hybrid orbitals and the unhybridized p orbitals shown in Figure 1.9(a) and (b) make it clear that unhybridized p orbitals on atoms joined by a sigma bond cannot point at each other. At best they can be parallel. These two parallel unhybridized p orbitals, one on each of the two atoms already connected by a sigma bond, can also overlap as shown in Figure 1.10. This type of overlap is called *pi (π)* overlap, and the bonds that result are called *pi bonds*. Pi overlap is not nearly as effective as sigma overlap, but it is effective enough to form a second and even a third bond between the two atoms.

(a)

(b)

Figure 1.9

(a) sp^2 hybrid orbitals and unhybridized p orbital; (b) sp hybrid orbitals and unhybridized p orbitals

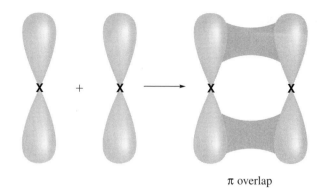

Figure 1.10

Pi overlap between unhybridized p orbitals

π overlap

This σ–π picture of a multiple bond is a very useful one. It helps us explain many observed chemical and physical properties of multiple bonds. But it is a picture that we have created to help us explain these properties. If you could look at a molecule with a multiple bond, you would not see σ bonds and π bonds. All you would see is, a cylindrical electron cloud between the two bonded atoms.

Think of some things that you can infer from the σ–π picture. If the π bond is weaker than the σ bond, a double bond should not be twice as strong as a single bond. Nor should a triple bond be three times as strong as a single bond or 1.5 times as strong as a double bond. Indeed they are not, as we shall see. If the electrons in the π bond, which are called π electrons, are not as tightly held between the two atoms, compounds with double bonds should react more readily with positively charged species seeking electrons, than would compounds with only single bonds. Compounds with multiple bonds are indeed generally more reactive than those without them.

When two atoms in a molecule are joined only by a single bond, it is possible for those two atoms to rotate freely around the single bond because the σ overlap is not affected (Figure 1.11). You can picture the σ bond as a cylinder of electrons. But if those two atoms are also joined by a π bond, rotation around the single bond will change the parallel relationship of the unhybridized p orbitals and break the π bond. There is a great deal of experimental evidence that verifies free rotation around single bonds and restricted rotation around multiple bonds.

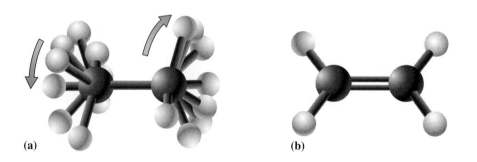

(a)　　　　　　　　　　　　　　(b)

Figure 1.11

Free rotation around σ bond; restricted rotation around π bond

Lone Pairs

In the examples we have looked at so far, the central hybridized atom is C. In virtually all its compounds, C uses all its valence electrons to make a total of four bonds and has no lone pairs remaining. But the other common elements in organic compounds, those in Groups VA, VIA, and VIIA, almost always form fewer than four bonds and thus have one or more lone pairs. In determining the hybridization of a central atom with lone pairs, each lone pair is counted as if it were a σ bond. So the N in NH_3, which has three σ bonds and one lone pair, is sp^3 hybridized. Similarly the O in H_2O, which has two σ bonds and two lone pairs, is sp^3 hybridized. But the halogens, which have three lone pairs but form only one σ bond in most organic compounds, are unhybridized.

Example 1.8

Identify the orbitals that overlap to form the bonds and that hold the lone pairs in the following molecules.

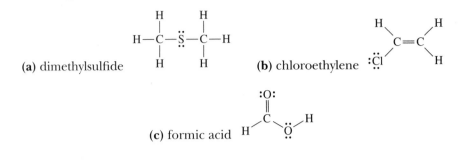

(a) dimethylsulfide

(b) chloroethylene

(c) formic acid

Solution

To determine the hybridization of an atom, add the number of atoms to which it is bonded and the number of lone pairs (if any). If the sum is 4, the atom is sp^3 hybridized. If it is 3, the atom is sp^2 hybridized, and if it is 2, the atom is sp hybridized.

But if the atom is bonded to only one other atom, we describe it as being unhybridized; it just uses ordinary atomic orbitals. Sometimes you will encounter an orbital picture in which an atom bonded to one other atom by a double bond or a triple bond is described as hybridized. For example, the O in CH_2O (Section 1.2) can be described as sp^2 hybridized, or the N in HCN can be described as sp hybridized. Either description can be used, but neither is totally satisfactory. What we call the "unhybridized" atom still has some mixing of orbitals compared to an isolated atom because it is bonded to something. What we call the "hybridized" atom, which uses only one hybrid orbital for bonding, has less mixing than an atom that uses at least two hybrid orbitals for bonding.

(a) Each C is bonded to four other atoms, so each is sp^3 hybridized. The S is bonded to two other atoms and has two lone pairs. The sum is four and the S is $3sp^3$ hybridized. Each H is bonded to one other atom, so it is unhybridized and uses its 1s orbital to form bonds. These hybridizations are shown in the diagram below:

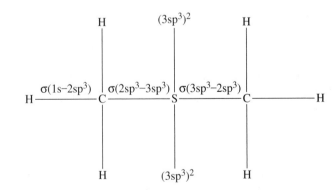

All C—H bonds are the same

(b) Each C is bonded to three other atoms, so it is sp^2 hybridized. The C atoms are joined by a double bond, which consists of a σ bond formed by overlap of two sp^2 orbitals and a π bond formed by overlap of two parallel p orbitals, for example, a $2p_x$ of each C. Each H is bonded to one other atom, so it is unhybridized and uses its 1s orbital to form bonds. The Cl is also bonded to only one other atom, so it is unhybridized and uses one of its 3p orbitals to form a σ bond with the sp^2 of carbon. Since the Cl is in the same plane as the other atoms bonded to this carbon, the 3p orbital it uses must be perpendicular to the $2p_x$ orbitals used by the C atoms to make a π bond between them. It can be either the $3p_y$ or the $3p_z$. These hybridizations are shown in the diagram below.

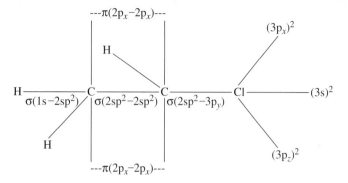

All C—H bonds are the same

(c) The C is bonded to three other atoms, so it is sp^2 hybridized. The H is joined by overlap of its 1s orbital with an sp^2 orbital of the C. The O joined to the C by a single bond is bonded to two atoms and has two lone pairs. The sum of four indicates that the O is sp^3 hybridized. Its σ bond to the C is formed by the overlap of its sp^3 orbital with an sp^2 orbital of the C. Its σ bond to H is formed by overlap of another of its sp^3 orbitals with the 1s orbital of the H. Its two lone pairs are in the remaining sp^3 orbitals. The other O is unhybridized because it is bonded to only one other atom, the C. It bonds to the C using one of its 2p orbitals to form a σ bond with an sp^2 orbital of the C. It uses another one of its 2p orbitals to form a π bond with the C. The two remaining lone pairs are in the other 2p orbital and the 2s orbital. These hybridizations are shown in the diagram below:

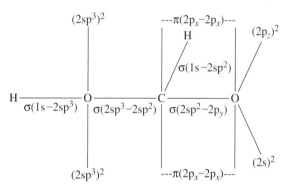

Notice that we could also have described this O as sp^2 hybridized. In this description, the O uses an sp^2 hybrid orbital to form the σ bond with the sp^2 of the C and uses its unhybridized 2p orbital to form the π bond with the C. The two lone pairs are each in an sp^2 hybrid orbital.

Neither of these two descriptions of the orbitals of an atom bonded to only one other atoms is satisfactory. The orbitals of this atom are not completely unhybridized because the overlap demands of even just one σ bond lead to some mixing of orbitals. But, we determine hybridization from measured bond angles, and there is no bond angle.

1.6 | Molecular Geometry

A description of the geometry of a molecule is a description of the positions of all the atoms in the molecule relative to each other. There are two critical parameters that we use for this description: bond angles and bond lengths.

To understand these parameters, we go back to Lewis Structures. Two atoms are joined by a line. When a given atom is joined to more than one other atom, these lines intersect at the given atom, which is called a central atom, and form angles

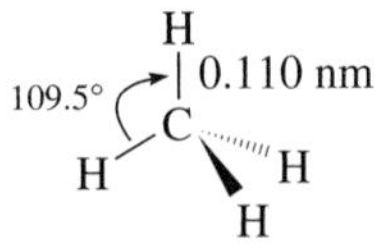

Figure 1.12
Geometry of methane

with each other. These angles are called **bond angles** and are shown in Figure 1.12 for methane, CH_4, in which the central atom is the C.

The line or bond joining two atoms in a Lewis structure is understood to be connecting the nuclei of the two atoms. The length of the line, or in other words the distance between those two nuclei, is called the **bond length**. By specifying the bond angles and bond lengths of a molecule, we are providing a description of its geometry. Figure 1.12 also shows the bond lengths in methane. You can see that the information provided by the bond angles and the bond lengths gives a complete description of the positions of the atoms in methane relative to each other.

Lewis structures are two-dimensional representations of molecules and do not even attempt to describe molecular geometry. The Lewis structure of methane appears to suggest that methane is flat and has bond angles of 90°. But because molecules are three dimensional, a picture of molecular geometry must also convey information about the third dimension. Since our drawing surface is flat, we use a projection, which is a two-dimensional drawing that includes information about the third dimension. Organic chemistry uses various kinds of projections depending on the three-dimensional information to be conveyed.

One simple projection that is used in Figure 1.12 is called the dashed line-wedge projection. Bonds are represented as ordinary lines if they are in the plane of the paper. They are represented as solid wedges if they are pointing toward front and up from the paper and as dashed wedges if they are pointing toward back and down from the paper.

One of the main reasons the hybridization model is so important is that it allows us to predict and to describe molecular geometry, even of molecules much more complex than methane. Let us start by looking at bond angles and the resulting molecular shapes, which are closely related to bond angles.

Ideal Geometries

As we have seen, each of the three kinds of hybrid orbitals that can be formed by combinations of the 2s orbital and one or more of the 2p orbitals has an associated **ideal** geometry of the substituents on the central atom. The four sp^3 orbitals correspond to tetrahedral geometry and lead to bond angles of 109.5° 28'. The three sp^2 orbitals correspond to trigonal planar geometry and lead to bond angles of 120°. The two sp orbitals correspond to linear geometry and lead to bond angles of 180°.

When an atom is sp^3 hybridized, the four atoms to which it is bonded will lie at or close to the four corners of a regular tetrahedron. If the four atoms are identical (such as the four H atoms of methane, CH_4, or the four Cl atoms of carbon tetrachloride, CCl_4), then they are exactly at the four corners and the molecule has ideal tetrahedral geometry. If the four atoms are not identical, the geometry will not be ideal. But it will often be very close, and we will still describe the geometry as tetrahedral. By far the most common atom found bonded to C in organic compounds is H, and the second most common atom is another C. When a C atom is bonded only to H atoms or other C atoms, its geometry will usually be very close to ideal tetrahedral if it has four σ bonds. In ethane, C_2H_6, all the HCH bond angles and all the HCC bond angles are considered to be 109.5° 28'.

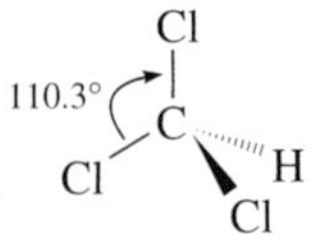

The extent to which the geometry deviates from the ideal will depend on the nature of the four atoms and occasionally on some other factors as well. Figure 1.13 shows the molecular geometry of chloroform, $CHCl_3$, and methylene chloride, CH_2Cl_2, both of which are tetrahedral, but not ideal. We will discuss ways in which we can predict the deviations from the ideal.

When an atom is sp^2 hybridized, the three atoms to which it is bonded will lie in a plane. If the three atoms are identical and the three bonds are identical, then the atoms lie at the corners of an equilateral triangle, the bond angles are 120° and the geometry is ideal. But when carbon has a completed octet, it forms four bonds, so one of the three atoms must be connected by a double bond in a neutral species. We find ideal trigonal planar geometry in a cation such as CH_3^+, in which carbon does not have a completed octet. We also find ideal trigonal planar geometry in

(a) Chloroform

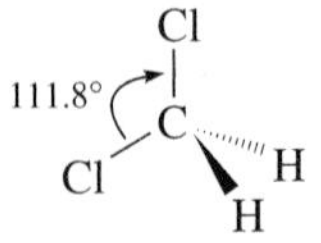

(b) Methylene chloride

Figure 1.13
Geometry of (a) $CHCl_3$ and (b) CH_2Cl_2

compounds where Group IIIA elements, which have incomplete octets, are bonded to three identical atoms. An example is BF_3. When a C atom is bonded only to H and to other C atoms by single or double bonds, its geometry will be close to ideal trigonal planar, and we shall consider the bond angles in a compound such as ethene, C_2H_4, to be 120° even though they deviate slightly.

When an atom is sp hybridized, the geometry is almost always linear and the bond angles are 180°.

Nonideal Geometries

The ability to predict and understand the geometry of a molecule even when it is nonideal is a very useful skill for organic chemistry. Our approach will be to look at every atom in a molecule and assign it a hybridization based on the number of σ bonds and lone pairs it has. For this purpose we will describe atoms with only one σ bond as unhybridized. We will not describe an atom with one σ bond as hybridized, since it has no associated bond angle and is not a central atom. An atom with a total of four σ bonds and lone pairs is sp^3 hybridized. An atom with a total of three σ bonds and lone pairs is sp^2 hybridized. An atom with two σ bonds and no lone pairs is sp hybridized. In Figure 1.14, we assign a hybridization to every central atom in acrylic acid.

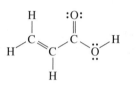

Once we assign a hybridization to a central atom, we can assign it the appropriate ideal geometry. The ideal geometry is our starting point for approaching nonideal geometry.

Deviations from ideal geometry occur because the substituents on a given central atom are not all the same, as shown in Figure 1.13. We are able to predict the effect of those differences, using two different models.

One model called the valence shell electron pair repulsion (VSEPR) model focuses on valence electrons and how they are best arranged. The other model focuses on the hybrid orbitals and how they change with different bonding or

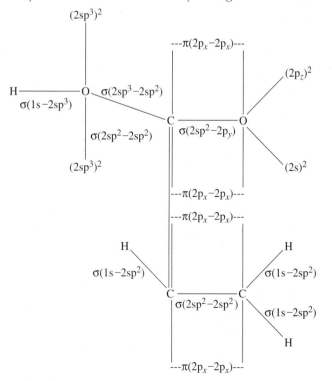

Figure 1.14
Acrylic acid, hybridization

valence electron situations. Both of these models seek to identify the geometry that is of lowest energy. Depending on the molecule, one model may be easier to apply than the other. We need to be able to use both. But, since both of these models focus narrowly on only one aspect of chemical bonding, they cannot always explain all the experimental results.

The VSEPR Model for Nonideal Geometry

The VSEPR model explains the observed geometry of molecules by an electrostatic picture based primarily on the Lewis structure. The model assumes that each electron pair in the valence shell of an atom in a molecule occupies a well-defined region of space. The first and most important rule of VSEPR is that the bond angles about an atom are those that minimize the total repulsion between the electron pairs in the valence shell of the atom.

To apply the VSEPR model to a molecule, we determine for each central atom the number of atoms and the number of lone pairs around it. This total is considered to be the number of valence shell electron pairs. Note that the four electrons of a double bond or the six electrons of a triple bond count as only one pair. There is an ideal geometry associated with the number of valence shell electron pairs, just as there is an ideal geometry associated with the number of hybrid orbitals.

Calculations show that repulsions between four electron pairs around an atom are minimized when the atoms bonded to it lie at the corners of a regular tetrahedron. For three electron pairs, they lie at the corners of an equilateral triangle. For two electron pairs, they lie on the same line. In other words the VSEPR model predicts exactly the same ideal geometry as the hybridization model.

To predict nonideal geometry, the VSEPR model considers each valence shell electron pair and corrects the bond angles of the ideal geometry by minimizing repulsions based on where the pair is located and/or the size of the pair. These corrections are done according to a number of rules that are part of the VSEPR model.

1. A lone pair is larger than a bonded pair.

   When there is a lone pair on an atom, the bonded pairs move away from the ideal, becoming further from the lone pair and therefore closer to each other. For example, in NH_3, Figure 1.15(a), the hybridization is sp^3 because the total of bonds and lone pairs is four. The geometry is not ideal, however, because of the relatively large size of the lone pair. Moving the bonded pairs away from the lone pair results in smaller bond angles. The H—N—H bond angle is 107°.

2. A multiple bond is bigger than a single bond.

   Electron pairs in single bonds tend to move further from the pairs in a multiple bond and closer to each other. In ethylene, C_2H_4, Figure 1.15(b), the hybridization is sp^2, but the H—C—H bond angle is only 118°.

3. An electron pair in a bond to a relatively electronegative atom is further away from the central atom.

   Because the electron pair in such a bond is further away from the central atom, the other bonds can get closer to it and further away from each other. In methyl chloride, Figure 1.15(c), the hybridization is sp^3, but the H—C—H angle is 110.9° and the H—C—Cl angle is 108°.

The Hybridization Model for Nonideal Geometry

In the hybridization model we can modify hybrid orbitals to better accommodate lone pairs or bonds to different kinds of atoms. To modify a hybrid orbital means to change its fractions of s orbital and p orbital, what we call its s character and p character.

You can see from the bond angles of the ideal hybridizations that the greater the s character the larger the bond angles. For sp^3 hybridization, the fraction of s is 0.25 and the ideal bond angles are 109.5°, while for sp^2 hybridization, the fraction

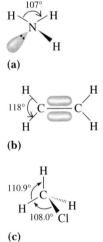

Figure 1.15
Bond angles in (a) NH_3, (b) C_2H_4, (c) CH_3Cl

of s is 0.33 and the ideal bond angles are 120°. For sp hybridization, the fraction of s is 0.50 and the ideal bond angles are 180°.

A consideration of the shape and size of an s orbital and a p orbital with the same principal quantum number indicates that there is more electron density near the nucleus in an s orbital than in a p orbital. This observation is also consistent with the lower energy of electrons in an s orbital compared to those in a p orbital.

We can understand nonideal geometry by considering the s character expected for a given hybrid orbital. The fraction of s will depend on the function of the hybrid orbital.

Consider the hybrid orbital in which a lone pair is found. Unlike a bonding pair of electrons, which has favorable electrostatic interactions with the two bonded atoms (Section 1.1), a lone pair interacts with only one nucleus. It is therefore best to have it as close as possible to that nucleus. While still maintaining hybrid orbitals suitable for bonding, the hybrid orbital of the lone pair will have more s character than the ideal, so it can get closer to the nucleus. As a consequence the hybrid orbitals used to form bonds will have less s character than the ideal. Less s character means a smaller bond angle, which explains the smaller than ideal bond angles around atoms with lone pairs, such as the N in NH_3 or the O in H_2O, which has an H—O—H bond angle of 104.5°.

Similarly, the s character of a hybrid orbital will depend on the atoms that are bonded. Consider the difference between a C bonded to an H and a C bonded to a more electronegative atom such as a Cl. The electronegative atom attracts the electron pair in the bond and moves it further from the C nucleus. Hybrid orbitals with less s character are used for such bonds because they have less electron density close to the C. The extra s character that results in the other orbitals can bring other electrons closer to the nucleus. Again, the orbitals with less s character will have smaller-than-ideal bond angles, explaining the smaller-than-ideal H—C—Cl bond angle of 108° in CH_3Cl.

The s character of a hybrid orbital sometimes reveals important information, not just about the geometry of a molecule but also about some of its properties. Differences in behavior between atoms that are sp^3 or sp^2 or sp hybridized can often be understood on the basis of the s character of the hybrid orbital, as we shall see.

Bond Lengths

Bond length is another very important parameter of molecular geometry. As we have seen, the line or bond joining two atoms in a Lewis structure is understood to be connecting the nuclei of the two atoms. The length of the line, or in other words, the distance between those two nuclei, is called the bond length. A number of factors influence bond lengths.

One critical factor is the size of the two atoms of the bond. Larger atoms form longer bonds. One important consequence of this rule is that a bond between a given atom and H, by far the smallest atom, is always shorter than a bond between that atom and anything else. For example, C—H bonds are always shorter than C—C bonds.

Besides being intuitively obvious, we can also explain this rule by considering the orbitals that overlap to form the bond. The size of an orbital increases with an increase in its principal quantum number. For example, a C—Cl bond, in which the 3p orbital of Cl overlaps with an sp^3 orbital of C, is shorter than a C—Br bond, in which the 4p orbital of Br overlaps with an sp^3 orbital of C.

Even when orbitals of similar size overlap, bond lengths can vary significantly. Bond order, which is generally the number of electron pairs shared between the two bonded atoms, can affect bond length in an important way. The greater the bond order, the shorter the bond. A triple bond is shorter than a double bond, which is shorter than a single bond. For example, the C≡C bond length in C_2H_2 is 0.120 nm, while the C=C bond length in C_2H_4 is 0.134 nm, and the C—C bond length in C_2H_6 is 0.154 nm.

The length of a bond of a given bond order also depends on the fraction of s character in the hybrid orbital forming the bond. A larger fraction of s character means a shorter bond. Single bonds of sp³ orbitals (0.25 s) are longer than single bonds of sp² orbitals (0.33 s). These single sp² bonds are longer than single bonds of sp orbitals (0.50 s). This hybridization effect explains the observed trend in C—H bond lengths. The C—H bond length in C_2H_6 is 0.110 nm, while the C—H bond length in C_2H_4 is 0.107 nm, and the C—H bond length in C_2H_2 is 0.106 nm.

A more subtle effect of atomic size on bond lengths can be seen in going from left to right in a row of the periodic table. Atomic size decreases as we go from C to N to O to F. The length of a bond between an sp³ C and each of these atoms also decreases. The C—F bond is shortest. Next is the C—O bond, then the C—N bond, and finally the C—C bond is the longest. Although an important factor, atomic size is not the only reason for this trend.

1.7 | Polarity of Bonds and Molecules

Homonuclear diatomic molecules—molecules consisting of two identical atoms—have completely covalent bonds; the two atoms share the bonded electrons equally. But electron sharing in all other bonds is not completely equal. The bonding electron pair or pairs are more closely associated with one of the two bonded atoms. As a result, each of the two atoms that are bonded has a partial electrical charge. We call this pair of charges a **dipole**. The atom with the greater share of the bonding electrons has a relative excess of negative charge, while the atom with the lesser share has a relative excess of positive charge.

A property of atoms called **electronegativity** has been devised that enables us to predict at least roughly the extent to which electron sharing between two given atoms is unequal. The electronegativity of an atom is a measure of the relative ability of that atom to attract shared electrons to itself when bonded to other atoms. Generally, the greater the difference in the electronegativity between the two atoms of a bond, the more unequal is the electron sharing. The range of electronegativity values of the elements found in organic molecules is relatively small because they are all nonmetals. These values are shown in Table 1.2.

Bonds can also be classified according to the degree of unequal electron sharing or the magnitude of the resulting dipole. Bonds in which electron sharing is equal or almost equal are called **nonpolar bonds**. Bonds with unequal electron sharing are called **polar bonds**. Bonds with almost no electron sharing are called **ionic bonds**. The bonds in organic molecules are either nonpolar or polar, but only very rarely ionic.

Bonds between carbon and hydrogen are nonpolar bonds; they have a very small dipole. Molecules that contain only carbon and hydrogen, called hydrocarbons, are nonpolar. Bonds between carbon and any of the other nonmetals are polar bonds. Organic molecules that contain elements other than just C and H not only have polar bonds but tend to have an overall polarity. Such molecules are polar, except in certain cases where the polarity of the individual bonds cancels, and the molecule as a whole is nonpolar.

Many physical properties of molecules are directly related to whether they are nonpolar or polar. Nonpolar compounds can be gases, liquids, or solids at room temperature. The solids tend to have low melting points and are soft and waxy. They dissolve in nonpolar organic solvents, such as gasoline, kerosene, or other liquid hydrocarbons. Polar compounds are usually liquids or soft solids at room temperature. They dissolve in polar solvents and to a lesser degree in some nonpolar solvents.

The overall polarity of a molecule depends not only on the polarity of its bonds but also on its geometry. Electrical polarity is expressed as a **dipole moment**, which is a measure of the magnitude of the separated charges and the distance between them. Each polar bond has a dipole moment, usually called a bond moment. A molecule has an overall molecular dipole moment, which is approximately the sum of all its individual bond moments. This sum, which is a vector sum, must include not only the magnitude of the bond moments but also their direction.

Using this description, we can understand why some molecules that have atoms of different electronegativities have no molecular dipole moments. Although they have polar bonds, the bond moments are oriented in such a way that they cancel

Table 1.2		Electronegativities of Some Nonmetals			
H 2.2	C 2.6	N 3.0	O 3.4	F 4.0
P 2.2	S 2.6	Cl 3.2	Br 3.0	I 2.7

each other. We represent bond moments by arrows pointing toward the negative end of the dipole. This representation helps us understand the relationship between the molecular geometry and the molecular dipole moment.

For example, each of the two double bonds in CO_2 is polar because O is much more electronegative than C, but CO_2 is nonpolar overall. Each of the two bond moments points toward an O.

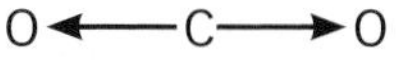

$$O \longleftarrow C \longrightarrow O$$

But CO_2 is linear, so the bond moments, which are equal in magnitude, point in exactly opposite directions and cancel each other.

In general, molecules that have identical bond moments symmetrically arranged will not have overall dipole moments. Compounds such as BF_3, which have three identical bond moments that point toward the three corners of an equilateral triangle (Figure 1.16), are examples of such molecules. The bond moments cancel each other.

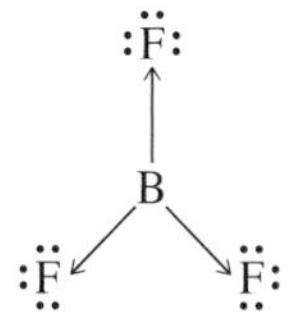

Molecules in which a single carbon atom has four identical substituents always have overall dipole moments of zero. Four identical bond moments that point toward the four corners of a regular tetrahedron cancel each other. Carbon tetrachloride, CCl_4, is a well-known example of such a molecule. Its bonds are quite polar, but its overall dipole moment is zero (Figure 1.17).

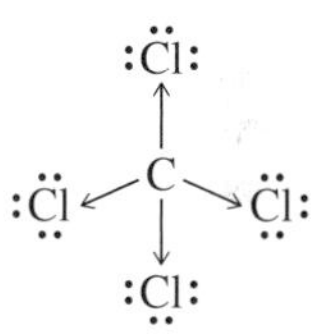

Although there appears to be a small difference in electronegativity between C and H, hydrocarbons are generally considered not to have dipole moments of consequence. The molecules are nonpolar.

Many chemical properties of molecules can be understood by considering the polarity of the molecule and its individual bonds. The polarity of a bond in a molecule can depend not only on the electronegativity difference between the two bonded atoms but also on the structure of nearby parts of the molecule. For example, the C—H bond in a molecule such as chloroform, $CHCl_3$, is more polar than an ordinary C—H bond because the three highly electronegative Cl atoms are pulling electron density toward themselves and away from the H (Figure 1.18).

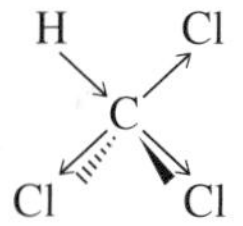

A well-known example of this effect is the increase in the acidity of oxyacids, which contain O, H, and a third element, with an increase in the electronegativity of the third element. For example, Cl—O—H is a stronger acid than Br—O—H, which is a stronger acid than I—O—H, because of the increase in the polarity of the O—H bond with the increase in the electronegativity of the third element. The observed increase in acidity whenever there is electron withdrawal from the bond to the acidic proton is a more general manifestation of this effect.

The polarity of a bond can be influenced by the structure of other parts of the molecule in two ways. Influences due to the electronegativity of nearby atoms or the presence of a nearby charge are called inductive effects. These influences are transmitted through the σ bonds of the molecule. Influences due to the distribution of the π electrons are called resonance effects. They can be identified by considering the electron distribution in the contributing structures of the resonance hybrid.

Figure 1.16
BF_3 bond moments

Figure 1.17
CCl_4 bond moments

Figure 1.18
Polarity of C—H bond in chloroform

1.8 | Bond Strength

Since every chemical change must include making or breaking at least one bond, the energy change associated with such a process is one key element in understanding and predicting chemical changes. The strength of a bond is measured by the quantity of energy that is released when the bond forms or the quantity of energy that is required to break the bond. This quantity is called the **bond energy,** and it is defined specifically for making or breaking bonds in the gas phase. It is expressed as a positive number in units of kJ/mol. The larger the number, the stronger the bond.

bond energy
Organic chemists use the term energy for this quantity, although it is actually enthalpy.

The bond energy of a bond depends not only on the identity of the two atoms that are bonded but also to some extent on the structure of the molecule of which they are a part. We can make some rough predictions about the relative strength of bonds using the hybridization model. Since bonds form by orbital overlap, we predict that the better the overlap, the stronger the bond.

Rather than measure and tabulate the bond energy of innumerable different bonds between the same two atoms, we instead define an *average* bond energy, which is the result of many measurements of different bonds between these two atoms. It is a characteristic of the two bonded atoms and the bond type. The average bond energy is usually a good approximation of the actual bond energy of a given bond. Table 1.3 lists the average bond energies of some bonds commonly found in organic molecules.

Even from the limited amount of data in Table 1.3, it is possible to identify some factors that influence bond strength and are consistent with orbital overlap.

A triple bond is stronger than a double bond, which is stronger than a single bond, between the same two atoms. For carbon–carbon bonds the numbers are 812 kJ/mol, 615 kJ/mol, and 346 kJ/mol. This trend is consistent with the number of orbitals that overlap to form each bond. Six orbitals (two sp and four p orbitals) overlap to form a triple bond. Four orbitals (two sp^2 and two p orbitals) overlap to form a double bond, while only two sp^3 orbitals overlap to form a single bond.

Looking more closely at the data reveals another connection between orbital overlap and bond energy. The bond energy of a C—C double bond is not twice that of a C—C single bond. While both types of bonds have a pair of hybrid orbitals with sigma overlap, the second pair of orbitals in the double bond are unhybridized p orbitals with pi overlap. Since pi overlap is weaker than sigma overlap, the pi bond is not as strong as the sigma bond. Similarly, a C—C triple bond is not 3 times stronger than a C—C single bond, nor 3/2 times stronger than a C—C double bond. The third bond of the triple bond is also a relatively weak pi bond.

The data in Table 1.3 also reveal a connection between bond length and bond strength. Orbital overlap becomes less effective as bonds get longer. The bond energies of bonds between C and a halogen—C—F 489 kJ/mol, C—Cl 328 kJ/mol, C—Br 285 kJ/mol, and C—I 218 kJ/mol—decrease as the size of the halogen increases and as the lengths of the bonds increase. The effect of small size is also seen in the relatively strong bonds that H forms. For example, the C—C bond energy is 346 kJ/mol, while the C—H bond energy is 413 kJ/mol.

Still another effect on bond energy is the difference in electronegativity between the two bonded atoms. The greater the difference, the stronger the bond. An example is C—N 305 kJ/mol, C—O 358 kJ/mol, and C—F 489 kJ/mol. This trend parallels the increase in electronegativity from N to O to F. The same trend is seen in the bond energies of double bonds, C=N 615 kJ/mol and C=O 749 kJ/mol.

Table 1.3	Average Bond Energies (kJ/mol)		
Bond	**Energy**	**Bond**	**Energy**
H—C	413	C—P	264
H—N	391	C—S	272
H—O	463	C=C	615
C—C	346	C=N	615
C—N	305	C=O	749
C—O	358	C≡C	812
C—F	489	C≡N	890
C—Cl	328		
C—Br	285		
C—I	218		

Chemical changes are the result of breaking of bonds in the reactants and making of bonds to form products. The heat of a reaction is the difference between the energy that is required to break the bonds and the energy that is released when the bonds are made. When energy is absorbed by a system, its sign is positive. When energy is released by a system, its sign is negative. The relationship between bond energies and the heat of the reaction is therefore ΔH(reaction) = Σ energies of bonds broken − Σ energies of bonds made. We shall see that the analysis of reactions in terms of bonds broken and bonds made is useful not only for calculating the heat of a reaction but also for understanding different aspects of chemical behavior.

Exercises

1.1 Indicate which of the following species have at least one atom that does not obey the octet rule in the given species.
(a) CO; (b) NO; (c) NCl_3; (d) ClF_3; (e) HNO_3; (f) H_3PO_4; (g) CBr_3^-; (h) AlF_3

1.2 Identify the atom that does not obey the octet rule in each of the indicated species in Exercise 1.1.

1.3 For elements 3 to 9 of the periodic table, formulate a rule that relates the preferred number of bonds to the number of valence electrons.

1.4 What is the relationship between the number of valence electrons in the p subshell of a nonmetal and its preferred number of bonds?

1.5 Write two different skeleton structures with the composition NH_3O and indicate which one corresponds to the more stable molecule.

1.6 The compound dimethyl ether has the formula C_2H_6O and does not have an O—H bond. Draw its Lewis structure. Draw the Lewis structure of ethyl alcohol, which also has the formula C_2H_6O but does have an O—H bond.

1.7 The compounds propyl alcohol and isopropyl alcohol both have the formula C_3H_8O. Use the result of Problem 1.6 to draw Lewis structures of these two isomers.

1.8 A compound with the formula $C_2H_6O_2$ has one O—H bond and no O—O bond. Draw its Lewis structure. Draw the Lewis structure of another compound with the same composition.

1.9 Draw the Lewis structures of two isomers of C_3H_6.

1.10 Phosgene, a poison gas used during World War I, has the composition $COCl_2$. Draw two different skeletal structures for this compound and indicate which is most likely to be correct.

1.11* Draw the Lewis structures of the two most stable isomers of CN_2H_4.

1.12 Benzene, C_6H_6, has a cyclic structure in which the six carbon atoms are joined together to form a regular hexagon, with a C atom at each of its corners. Each C is bonded to only one H. Draw the two most important contributing structures of benzene.

1.13* Methyl formate is an isomer of acetic acid that does not have an O—H bond. It has two important contributing structures one of which is much more important than the other. Draw the two contributing structures.

1.14 Draw the most important contributing structures of the bicarbonate ion, HCO_3^-.

1.15 Draw the three most important contributing structures of the carbonate ion, CO_3^{2-}.

1.16* The compound diazomethane, CH_2N_2, is very unstable. It has two contributing structures, which have formal charges. The C is bonded to one N. Draw the two contributing structures.

1.17* The azide ion, N_3^-, is a linear and symmetrical ion. Draw three important contributing structures for this ion.

1.18 Acrylonitrile, C_3H_3N, is an important starting material in the manufacture of polymers. It has the skeleton structure C—C—C—N. Draw the three most important contributing structures for this compound.

*More challenging questions.

1.19* Consider the different cations with the formula $C_2H_2Cl^+$. Use the theory of resonance to predict which is the most stable of these cations and draw its most important contributing structure.

1.20 A molecule has the formula C_4H_4. It does not contain any double or triple bonds. All the carbons are identical, and all the hydrogens are identical. All octets are complete. Draw the Lewis structure of this compound.

1.21 Draw orbital diagrams of the type shown in Example 1.8 for the following species.
(a) Formaldehyde; (b) $C_2H_2Cl_2$; (c) methyl ammonium ion; (d) nitromethane; (e) CH_3^-

1.22 Draw orbital diagrams of the type shown in Example 1.8 for the following molecules.
(a) CH_5N; (b) CH_3N; (c) C_2H_3N

1.23 Draw orbital diagrams of the type shown in Example 1.8 for the most stable isomers of each of the following.
(a) CHOCl; (b) CHOI

1.24 Draw orbital diagrams of the type shown in Example 1.8 for two noncyclic isomers with the formula C_3H_4.

1.25 Draw orbital diagrams of the type shown in Example 1.8 for two noncyclic isomers with the formula C_2H_4O.

1.26 Some of the following triatomic species are linear. Draw orbital diagrams for those that are.
(a) CO_2; (b) BrCN; (c) HOCl; (d) NOCl

1.27 Indicate the ideal geometry associated with each of the following species.
(a) CH_2F_2; (b) NCl_3; (c) NCO^-; (d) H_3O^+; (e) $COCl_2$ (phosgene); (f) NH_2^+

1.28 Arrange the following species in order of increasing H—C—H bond angles (smallest one first)
(a) CH_4; (b) CH_3^-; (c) CH_3^+; (d) CH_2O ; (e) CH_2Cl^+

1.29* Carbon suboxide is a linear molecule with the formula C_3O_2. Draw an orbital diagram for this unusual molecule.

1.30 Some of the following species are planar. Draw orbital diagrams for those that are.
(a) CCl_3^-; (b) C_2H_2O; (c) C_2H_3Br; (d) CO_3^{2-}

1.31* Draw orbital diagrams of three compounds with the formula C_4H_4 that are planar. (Hint: Two are cyclic.)

1.32 Draw the structure of an isomer of C_2H_3NO that fits each of the following descriptions.
(a) Noncyclic, no N—O bond, all atoms except H and O with the same hybridization (two possibilities); (b) noncyclic, no N—O bond, each C with a different hybridization; (c) cyclic, an N—O single bond, both Cs with the same hybridization (two possibilities)

1.33 The bond angles in NF_3 are 102°, much smaller than the bond angles in NH_3, which are 107°. Explain the difference using a VSEPR argument and a hybridization argument.

1.34 Arrange these compounds in order of increasing length of their C—H bonds: C_2H_2, C_2H_4, C_2H_6.

1.35 Arrange these compounds in order of increasing length of their bonds between C and the element other than H: CH_5P, CH_4S, CH_3Cl, CH_3F.

1.36 Without referring to Table 1.3, list the following bonds in order of increasing strength: C—N, C—O, C—P, C—S.

1.37 Compare the bond energies of carbon–halogen bonds to those of bonds of carbon to other nonmetals to assess the relative importance of the factors determining bond strengths.

1.38 Arrange the following molecules in order of increasing dipole moment: CH_3Cl, CH_3SH, CH_3F, CH_3Br

1.39* Draw the structure of an isomer of $C_2H_2Cl_2$ that does not have a dipole moment.

*More challenging questions.

Chapter 2
Molecular Orbital Theory

Everything said so far about bonding has been based on two helpful simplifications. First, we have localized electron pairs, either as lone pairs on one particular atom or as bonded pairs between two particular atoms. Second, we have divided the valence electrons of a molecule into two groups, bonding and nonbonding, and assumed that they can be considered separately. Neither of these simplifications is correct. But without them, the description of the electronic structure of molecules with more than two atoms becomes extremely complex. Fortunately, such complex descriptions are usually not necessary in organic chemistry.

Nevertheless, there are many aspects of the behavior of organic molecules that cannot be understood simply on the basis of valence bond theory alone. It is for these situations that we abandon, at least to some extent, the two simplifications of valence bond theory and use some of the ideas of molecular orbital theory.

2.1 | Molecular Orbitals

We have already used the idea of combining two or more atomic orbitals of a given atom to make hybrid orbitals that better describe how the atom bonds to other atoms. In a similar way, we can describe a bond by making combinations of the atomic orbitals, unhybridized or hybridized, that overlap to form the bond between the two atoms.

The combining of atomic orbitals on different atoms must obey the same rules that we have already presented for making combinations of orbitals on the same atom to form hybrid orbitals.

1. The number of combination orbitals equals the number of atomic orbitals that are combined to make them. The number of orbitals associated with any chemical species remains constant no matter how we choose to combine or blend them.
2. The total energy of the combination orbitals is equal to the total energy of the atomic orbitals that are combined to make them. The energy of a chemical species does not change just because we choose to describe it differently.

When two atomic orbitals, one on each atom, overlap to form a bond between the two atoms, we can describe the process as the combining of the two orbitals. According to Rule 1, the combining of these two orbitals must produce two new orbitals. According to Rule 2, the total energy of these two new orbitals must be equal to the total energy of the atomic orbitals that combined to form them. The energy of one of the new orbitals is lower than the energy of the atomic orbitals. The energy of the other one is higher than the energy of the atomic orbitals by the same amount. These relationships are shown in Figure 2.1.

According to the Pauli Exclusion Principle, an orbital can hold a maximum of two electrons. The electron pair that is shared by the two bonded atoms is placed in the lower-energy orbital. This orbital is called the **bonding** molecular orbital. Its lower energy helps explain why the formation of bonds is energetically favorable. The higher-energy orbital can also hold up to two electrons. But because its energy

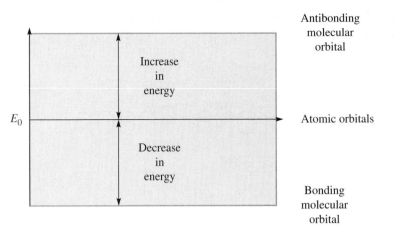

is relatively high, putting electrons into this orbital is energetically unfavorable and would result in destabilization. This orbital is called the **antibonding** molecular orbital. This picture helps us to understand why one bond is the sharing of one electron pair.

Consider the formation of H_2, in which the overlap is between a 1s orbital of each H. The two electrons of the molecule are accommodated in the bonding molecular orbital, and the molecule is stable. Now consider the formation of He_2, in which the overlap is between a 1s orbital of each He. This species has four electrons. Two are accommodated in the bonding molecular orbital, but the other two must be in the antibonding molecular orbital. The energy gain is exactly offset by the energy loss, and the He_2 molecule is not stable.

Double and triple bonds are easily explained using this picture. When two atoms are joined by a double bond, each atom uses two atomic orbitals. The combination of four atomic orbitals forms four molecular orbitals. Two of them are bonding, and two of them are antibonding. The four electrons of the double bond are in the two bonding molecular orbitals. The two bonding molecular orbitals have different energies, as do the two antibonding molecular orbitals. The same reasoning can be extended to a triple bond, which will have three bonding and three antibonding molecular orbitals.

The σ bond, which in the case of carbon is formed by overlap of two hybrid orbitals, is a stronger bond than the π bond, which is formed by the less effective overlap of two parallel unhybridized p orbitals (Section 1.5D). Because of Rule 2, the energy difference between the σ bonding molecular orbital and the hybrid orbitals is equal to the energy difference between the σ antibonding molecular orbital and the hybrid orbitals. The π bonding and antibonding molecular orbitals have the same energy relationships to the unhybridized p orbitals. Antibonding molecular orbitals are designated by the symbol *. The four orbitals associated with a double bond are called, in order of increasing energy, σ, π, π*, and σ*. Figure 2.2 describes the orbital energy relationships in molecules, such as ethylene, (C_2H_4) that have double bonds.

Although the lowest energy state of the double bond has no electrons in the antibonding molecular orbitals, the energies, shapes, and sizes of these orbitals can

Figure 2.2

Relative energies of σ, π, π*, and σ* orbitals

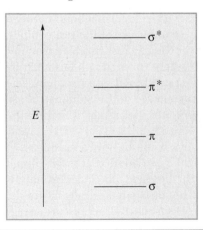

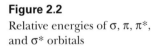

help us understand the chemical behavior of many substances. The reason is that when a molecule absorbs energy, either by interaction with radiation or as part of a chemical change, an electron moves from a bonding to an antibonding orbital.

Notice that the molecular orbitals that we are proposing here to describe the formation of a bond are not truly molecular orbitals, except in the case of diatomic molecules. They are associated with just two of the atoms of the molecule. This description is a simplification of the real molecular orbital description, which constructs orbitals associated with all the atoms of the molecule. Doing it the real way is extremely complicated for all but very simple molecules.

2.2 | Combinations of Atomic Orbitals

How is it possible for two atomic orbitals to combine in two different ways and form two orbitals of different energies? What do these two different orbitals look like?

We can answer these questions by focusing on the wave nature of an electron. There are two ways in which two waves can combine, constructively or destructively, as shown in Figure 2.3. If they combine constructively in phase (b), the resulting wave is the sum of the two waves. If they combine destructively out of phase (a), the resulting wave is the difference between the two waves.

When two atoms approach each other, their atomic orbitals, which are simply wave functions, can combine in phase constructively or out of phase destructively. We can show the phase of an atomic orbital with a + or − sign or with two different colors.

When the atomic orbitals combine constructively in phase, we indicate that by giving both orbitals the same sign (or the same color). The electron density between the two nuclei is increased, and a bonding orbital is formed. If the approach is along the line between the nuclei, the new orbital is a σ bonding molecular orbital. Figure 2.4 shows the constructive, in phase combination of the 1s orbitals of two H atoms to form the σ bonding molecular orbital of H_2, which is of lower energy than the atomic orbitals.

When the atomic orbitals combine destructively, out of phase, the orbitals have opposite signs as shown in Figure 2.4. The orbital that forms has a **node,** which is a region of no electron density. The node is between the two nuclei, and this orbital is of higher energy than the atomic orbitals. It is the σ* antibonding orbital. An s orbital has a spherical shape, and therefore the electron wave defined by this orbital has the same phase throughout the entire orbital. The orbital is represented with a single sign as shown in Figure 2.4.

Such is not the case for other atomic orbitals. More typical is a p orbital, which has two lobes and a node at the nucleus. The electron wave defined by a p orbital has different phases in the two lobes. Thus a p orbital is represented with a different sign or different color in each lobe.

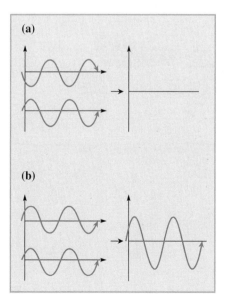

Figure 2.3

Combining two waves

Figure 2.4

Formation of the σ bonding molecular orbital of H_2 and the σ* antibonding orbital of H_2

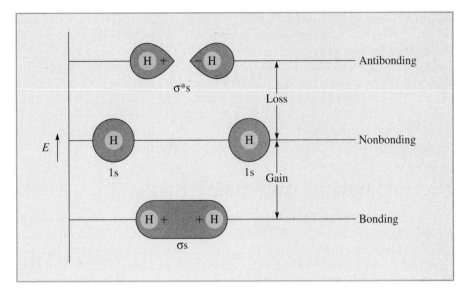

When a p orbital combines with any other orbital, the lobe that is in phase with the other orbital must be used to make the bonding orbital. The lobe that is out of phase is used to make the antibonding orbital. Figure 2.5 shows the formation of a σ orbital by combination of a 2p orbital with a 1s orbital. We would find such an orbital in H—F.

Two p orbitals can combine along the internuclear line to form a σ bond. When the combination is between lobes that are in phase (ones with the same sign or color), a σ bonding orbital is formed. When the combination is between lobes that are out of phase, a σ* antibonding orbital is formed. Figure 2.6 shows these possibilities. We can find such orbitals in F_2 where both atomic orbitals are 2p.

When two parallel p orbitals combine, the result is π bonding. Once again the two orbitals can combine constructively or destructively. As shown in Figure 2.7, if the phases of the two lobes of one of the p orbitals match the phases of the two lobes of the other, the combination is constructive and a π bonding orbital

Figure 2.5

Combination of a 2p and 1s orbital

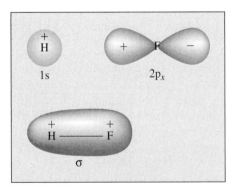

Figure 2.6

Combination of two 2p orbitals

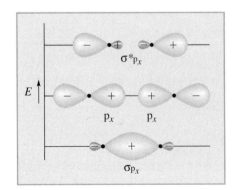

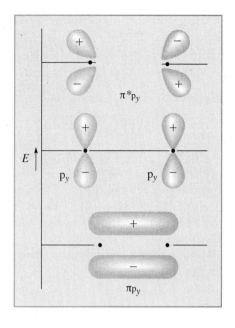

Figure 2.7
Combination of two 2p orbitals,
π bonding

results. If they don't match, the combination is destructive and a π* antibonding orbital results. Just as in the case of σ* orbital, there is a node between the atoms in the π* orbital. Just as in the case of σ bonding, the π bonding orbital is of lower energy than the atomic orbitals that combined and of much lower energy than the π* antibonding orbital. But the energy difference between bonding and antibonding π orbitals is not as great as the energy difference between bonding and antibonding σ orbitals. Because a σ bonding orbital is of lower energy than a π bonding orbital, the σ* orbital must be of higher energy than the π* orbital, since energy lost must equal energy gained. As we will see when we study compounds with double bonds, this smaller energy difference between π and π* orbitals can have an important effect on chemical behavior.

2.3 | Conjugation

As we've seen, the structures of many molecules cannot be described by a single Lewis structure but instead are described by blending two or more imaginary contributing structures. A modified molecular orbital description of such molecules often yields simpler and more informative pictures.

The most important characteristic of many molecules that are resonance hybrids is delocalization of electrons. In these molecules one or more electron pairs are distributed over three or more atoms rather than being localized between two bonded atoms. The contributing structures that we use to describe such molecules generally differ from each other in the arrangement of the π bonds. Therefore the molecular orbitals that we use to describe such molecules are those of the π bonds.

Molecular orbitals of π bonds are orbitals that form by π overlap of parallel p orbitals on adjacent atoms. We can construct these molecular orbitals by π overlap between one p orbital on each atom that takes part in multiple bonding in a contributing structure, provided these p orbitals are parallel to each other. Any species that has parallel p orbitals that can take part in bonding on three or more adjacent atoms is said to be **conjugated.**

The same rules that apply to combinations of two p orbitals on two adjacent atoms also apply to combinations of more than two p orbitals on more than two adjacent atoms.

1. The number of combination orbitals equals the number of atomic orbitals that are combined to make them.
2. The total energy of the combination orbitals is equal to the total energy of the atomic orbitals that are combined to make them.

Consider the cation represented by the two contributing structures:

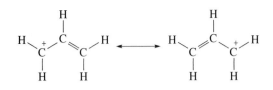

Each C atom has an unhybridized 2p orbital. These three 2p orbitals can overlap to form three molecular orbitals (Rule 1), each of which extends over the three atoms. One of these molecular orbitals is of lower energy than the atomic orbitals. It is the π bonding orbital. Another molecular orbital is of higher energy than the atomic orbitals by the same amount. It is the π* antibonding orbital. The third molecular orbital must then have the same energy as the atomic orbitals. It is called a nonbonding orbital. These energy relationships are shown in Figure 2.8.

Figure 2.8

Relative energy of the atomic orbitals and π molecular orbitals in a three-atom system

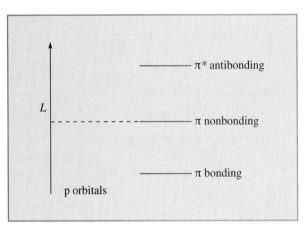

Notice that both the valence bond and the molecular orbital description give the same picture. We can describe the species as an equal blend between two equivalent contributing structures or as a species with a pair of π electrons in a bonding orbital spread over all three atoms. The electron pair is associated equally with the C atom at each end of the species.

These ideas can be extended to systems that have more than three adjacent atoms with parallel unhybridized p orbitals. One simple example of such a system is the compound 1,3-butadiene.

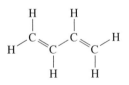

Since there are four unhybridized p orbitals, one on each C atom, and since they can be parallel to each other, we can form four molecular orbitals. Each of these orbitals occupies a region of space that includes all four atoms. Two of these orbitals are of lower energy than the atomic orbitals and are π bonding orbitals. The other two are of higher energy than the atomic orbitals and are π* antibonding orbitals.

Each of these molecular orbitals has a different energy. The lowest-energy bonding orbital is called the π_1 orbital, the next higher-energy bonding orbital is called the π_2 orbital. The two antibonding orbitals in order of increasing energy are the π_3^* and π_4^* orbitals. The energy relationships are shown in Figure 2.9.

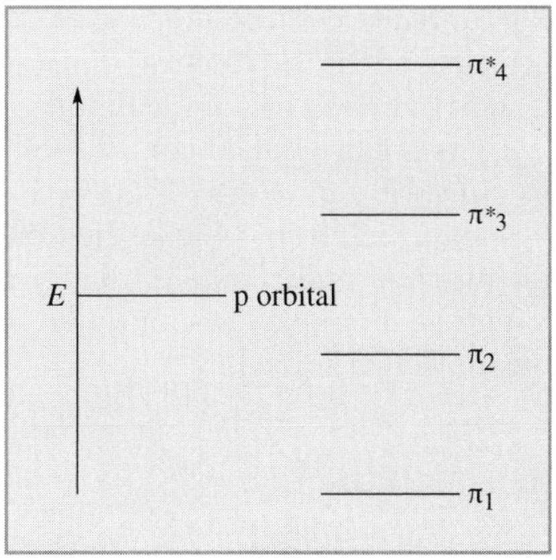

Figure 2.9
Relative energy of the atomic orbitals and molecular orbitals in a four-atom system

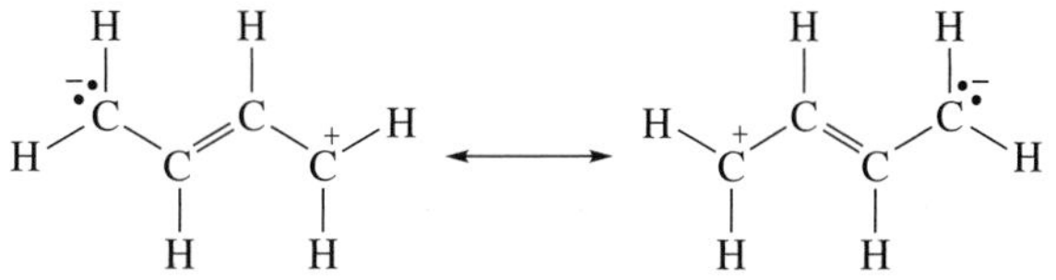

Figure 2.10
Contributing structures of 1,3-butadiene

Once again the molecular orbital description and the valence bond description are consistent. Two minor contributing structures (Figure 2.10) describe a molecule with four π electrons delocalized over four atoms with some double bond character between the two inside C atoms. Two bonding molecular orbitals spread over the four atoms convey the same picture. The sum of the energies of the π_1 and π_2 orbitals is lower than twice the energy of a π bonding orbital in a compound, such as ethylene, that has only one double bond.

$$\pi_1 + \pi_2 < 2\pi$$

The lower total energy of the two bonding orbitals is consistent with the resonance stabilization predicted by the valence bond description.

The energy relationships among the various molecular orbitals in conjugated systems provide a powerful basis for understanding a good deal of their chemical and physical behavior.

2.4 | Combinations of 2p Orbitals

The formation of two molecular orbitals of different energies from combinations of two atomic orbitals can be explained as the result of constructive or destructive overlap. We have already seen that when two p orbitals overlap to form a π orbital, they can do so in two ways (Figure 2.7). In constructive overlap, which leads to a more stable bonding orbital, the signs of the lobes of each p orbital match. In destructive overlap, which leads to the less stable antibonding orbital, the lobes do not match, and there is a node, a region of zero electron density between their two atoms. Thus it is the matching of the signs of each lobe that indicates the relative energy of the molecular orbital formed: low energy if they match; high energy if they do not.

The same approach can be used to evaluate the relative energies of orbitals that are formed by overlap of more than two p orbitals. We can represent these molecular orbitals schematically by drawing and indicating the orientation of the p orbitals that combine to form them. Such a picture is not a drawing of the actual size and shape of an orbital in the way that our drawings of atomic orbitals are. But such a picture is very helpful in presenting a coherent picture of the π molecular orbitals of a conjugated system and, as we shall see, in predicting and understanding various aspects of the behavior of these systems.

Once we assign signs and orientations to the p orbitals that are combining to form a molecular orbital, we can formulate a picture of that molecular orbital in the same way as we formulated a picture of the molecular orbitals in a compound with one double bond, such as ethylene, CH_2=CH_2. If the signs of the adjacent p orbitals match, the electron density will be between the two atoms. If they do not match, there will be a node between the two atoms (Figure 2.7).

The relative stability of a π orbital will be defined by the number of matches of its unhybridized p orbitals. As we have seen for the simple π* antibonding orbital formed from two p orbitals, the nonmatch results in a node between the nuclei of the two atoms. The same is true for systems in which more than two p orbitals combine. The more matches, or stated another way, the fewer nodes, the more stable is the molecular orbital.

For example, in 1,3-butadiene, in which there are four unhybridized p orbitals that combine, we can represent the π molecular orbitals schematically by drawing the four p orbitals with their signs and using the location of nodes to draw pictures of the four resulting molecular orbitals, as shown in Figure 2.11.

The lowest energy orbital, π_1, that forms is the one in which the signs of the lobes of all the p orbitals match and there are no nodes. It is continuously bonding between all four atoms. This orbital holds two π electrons with paired spins.

The orbital of the next lowest energy is the π_2, which is also bonding. It has electron density between C-1 and C-2 and between C-3 and C-4, but it has a node between C-2 and C-3. It is the energy equivalent of an isolated π bond. It also holds two π electrons with paired spins. The two antibonding orbitals, π_3* and π_4*, are unoccupied. In the least stable one, π_4*, there is a node, and therefore no bonding between any two adjacent atoms. In the other one, π_3*, there is bonding only between C-2 and C-3. In both of them there are more nodes between adjacent atoms than bonds. The relative energies of the four π molecular orbitals increase as the number of nodes (from nonmatches) increases.

Once a set of allowed molecular orbitals has been defined, the electrons originally in the atomic orbitals that combined can be placed in these orbitals, starting with the lowest-energy molecular orbital. As we know, no more than two electrons can be placed in an orbital. We can see from the structure of the molecule how many electrons need to be accommodated in these orbitals.

In 1,3-butadiene there are two double bonds and therefore four π electrons. Two of them fill the π_1 bonding orbital and the other two fill the π_2 bonding orbital. The compound is thus stable because all the electrons are in bonding molecular orbitals. This picture also reveals that two of the π electrons, the ones in the π_2 molecular orbital, are of somewhat higher energy than the two in the π_1 bonding

Figure 2.11

The four π orbitals of 1,3-butadiene (schematic and actual)

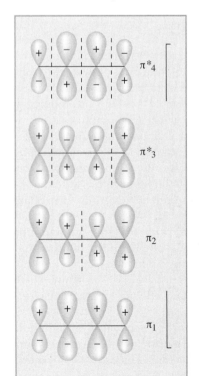

orbital. This picture will help us understand some of the chemical and spectroscopic properties of 1,3-butadiene and other similar conjugated systems.

2.5 | Electronic Transitions

The set of molecular orbitals that we construct by combinations of atomic orbitals of different atoms in a molecule provides a framework that we can use to describe the allowed energies and locations of its shared electrons. The allowed energies of an electron are the energies of the orbitals in which it may be found. The locations of the electron are in regions of space that are defined by the shape and location of the orbitals in which it may be found.

Remember an orbital is simply a description of a region of space in which an electron may likely be found if its energy is such that it is in that orbital. If you could look at a molecule, you would not see orbitals. You would see electron clouds in certain regions of space designated by the occupied orbitals. And you wouldn't see anything at all in the regions of space designated by unoccupied orbitals. In that case, what is the relevance of unoccupied orbitals, and in particular the antibonding orbitals?

As we know an atom can go from a lower energy state to a higher energy state, called an excited state, by absorbing radiation or from a higher energy state to a lower one by emitting radiation. Absorption of radiation is much more important than emission of radiation in organic chemistry. When energy is absorbed, an electron jumps to a higher-energy unoccupied orbital. The atom can only absorb radiation whose frequency corresponds to the energy difference between the lower-energy occupied orbital and the higher-energy unoccupied orbital. The relationship between the energy of the radiation and its frequency is given by

$$E = h\upsilon \quad \text{or} \quad \Delta E = h\upsilon$$

where h is Planck's constant and ΔE is the difference in energy between the two orbitals involved in the absorption of the particular radiation. An *electronic* transition is the change of an electron from one orbital to another that accompanies the absorption or emission of radiation.

Like atoms, molecules are also able to absorb radiation. In general, the radiation absorbed by organic molecules is in or near the ultraviolet/visible region of the electromagnetic spectrum. The frequency of the radiation corresponds to the difference in energy between an occupied orbital and an unoccupied one. In most cases the occupied orbital will be a bonding molecular orbital and the unoccupied one an antibonding molecular orbital. So knowing the identity of the various antibonding molecular orbitals is essential for determining what kinds of electronic transitions are possible in a given molecule. And knowing the energies of the antibonding orbitals, as well as the bonding orbitals, helps us to predict and understand quantitatively the absorption of radiation by a molecule.

The details of the electronic transitions that are allowed in a given molecule depend on the exact structure of the molecule. But we can identify at least the general kinds of possible electronic transitions and their relative energies by considering the types of electrons and orbitals in a molecule.

The only electrons we need to consider are the valence electrons of each of the atoms of a molecule. We can classify them as bonding electrons, which are in bonding molecular orbitals, or as lone pairs, which are in atomic orbitals, either hybridized or unhybridized. A molecule in its ground state, which is its lowest energy state, does not have electrons in antibonding orbitals. We can classify the bonding electrons as σ electrons, which are in σ bonding molecular orbitals, or as π electrons, which are in π bonding molecular orbitals. While all molecules have σ electrons, only molecules with π bonds (double or triple bonds) have π electrons.

Because σ bonds are stronger than π bonds, σ electrons are of lower energy than π electrons. Because bond formation always lowers the energy of the electrons that form the bond, lone pair electrons are of higher energy than even the π electrons. A lone pair electron is usually represented by the symbol n. Only molecules with atoms other than carbon and hydrogen have n electrons.

Figure 2.12

Relative electronic energy levels

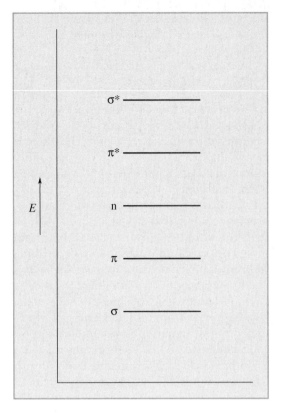

Any one of these three types of electrons can jump to a higher-energy vacant molecular orbital when the molecule absorbs radiation. Which electron jumps and where it jumps to depends on the energy/frequency of the absorbed radiation. Its energy must match the difference in energies (ΔE) between the orbital occupied by the electron and the higher-energy unoccupied orbital to which it jumps.

There are two types of higher-energy unoccupied orbitals, both are antibonding, the σ^* and the π^*. Because a σ bonding orbital is of lower energy than a π bonding orbital, a σ^* antibonding orbital is of higher energy than a π^* antibonding orbital. All of these energy relationships are shown in Figure 2.12.

This figure represents the rough energy relationships in a generalized molecule that has a π bond and an atom other than C and H. As we shall see, specific molecules may have simpler or more complex relationships. Based on this general picture, we can classify the possible types of electronic transitions that we may encounter. Since there are three types of electrons and two types of unoccupied orbitals, a total of six different kinds of electronic transitions are possible.

We use an arrow starting from the electron type and going to the unoccupied orbital to describe an electronic transition. The notation $\sigma \rightarrow \sigma^*$ is read as a "sigma-to-sigma star transition." It indicates a σ electron jumping to a σ^* antibonding orbital. As shown in Figure 2.12, it is the electronic transition with the greatest ΔE. It occurs when the molecule absorbs radiation whose frequency corresponds to this energy.

Since the energy and frequency of radiation are directly proportional ($E = h\upsilon$), it is the absorption of the highest frequencies of radiation that corresponds to the $\sigma \rightarrow \sigma^*$ transition. Frequency and wavelength of radiation are inversely proportional:

$$\upsilon = c/\lambda$$

where c is the speed of light and λ is the wavelength. So we can also say that it is the absorption of the shortest wavelengths of radiation that corresponds to the $\sigma \rightarrow \sigma^*$ transition.

The electronic transition with the smallest ΔE is the $n \rightarrow \pi^*$. It is accompanied by the absorption of the longest-wavelength radiation. There are four other

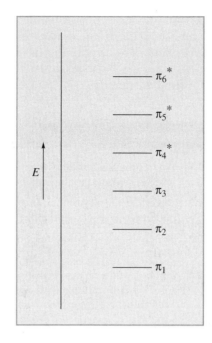

electronic transitions that are possible. Generally, the one with the next smallest ΔE is the $\pi \rightarrow \pi^*$. This kind of electronic transition is a very important one, especially in conjugated systems, as we shall see. The other three transitions are less important. They are the $n \rightarrow \sigma^*$, the $\pi \rightarrow \sigma^*$, and the $\sigma \rightarrow \pi^*$.

In general, electronic transitions involving σ electrons or σ^* orbitals are not so important. The reason is the relatively high ΔEs associated with transitions that involve the relatively low-energy σ electrons or the relatively high-energy σ^* antibonding orbitals. A high ΔE means that radiation of high frequency/short wavelength must be absorbed to bring about the transition. This radiation is of shorter wavelength than ultraviolet radiation. It is called vacuum ultraviolet radiation and is produced only under special conditions. Furthermore, when a molecule absorbs such high-energy radiation, it often undergoes decomposition. In particular, the excitation of a σ electron into an antibonding orbital generally breaks the σ bond and results in the formation of two fragments from the molecule.

The other electronic transitions, the $n \rightarrow \pi^*$ and the $\pi \rightarrow \pi^*$ in which π electrons or π^* orbitals are involved, have lower ΔEs and therefore take place with the absorption of longer-wavelength radiation. When a molecule has an occupied orbital relatively close in energy to an unoccupied one, the ΔE of the electronic transition is relatively small and the wavelength of the radiation is relatively long. Sometimes this radiation is in the ultraviolet or even the visible region of the electromagnetic spectrum. Such energy relationships among orbitals are found in conjugated molecules, which have relatively high-energy π bonding orbitals and relatively low-energy π^* antibonding orbitals.

Figure 2.9 shows the π and π^* orbitals in a four-atom conjugated system such as 1,3-butadiene. The π_2 bonding orbital and the π^*_3 antibonding orbital are much closer in energy than the π and π^* orbitals in ethylene and other compounds with unconjugated double bonds (Figure 2.2). A molecule such as 1,3-butadiene absorbs radiation in the ultraviolet region. Longer conjugated systems, which have more atoms, also have more π bonding orbitals and more π^* antibonding orbitals. The more of these orbitals there are, the smaller the ΔE between the highest-energy occupied π bonding orbital and the lowest-energy π^* antibonding orbital. Figure 2.13 shows the orbitals in a six-atom conjugated system.

Most molecules do not absorb radiation in the visible region of the electromagnetic spectrum. They are colorless. It is very easy to tell when a molecule does absorb visible light; it is colored. Molecules that absorb visible light tend to have long conjugated systems with occupied π bonding orbitals close in energy to unoccupied π^* antibonding orbitals.

An understanding of bonding and antibonding molecular orbitals helps us to understand electronic transitions and the interaction of molecules with ultraviolet and visible light. In addition, as we shall see, the details of how many chemical reactions occur can be better understood by considering the molecular orbitals that are involved in reactions.

Exercises

2.1 What is the total number of σ* orbitals in propene, C_3H_6? How many different energy levels correspond to these σ* orbitals?

2.2 What is the total number of σ* orbitals in cyclopropane, C_3H_6? How many different energy levels correspond to these σ* orbitals?

2.3 Sketch the σ and π bonding orbitals for the bonds between C and O in formaldehyde, CH_2O.

2.4 Under certain very special conditions it is possible to add an electron to ethylene to form a highly unstable species with a negative charge and an odd number of electrons, called a radical anion. Draw an orbital diagram of this species and use it to explain the instability of this species. Draw a Lewis structure of this species and use it to explain the instability of the species.

2.5 Consider the reaction between CH_3^+ and $H:^-$. As these two species approach each other, the pair of electrons of the $H:^-$ is donated to the CH_3^+. Identify the orbital on each of these species that is involved in this electron transfer. Identify the orbital in which the electron pair is found after the reaction is over.

2.6 Consider the reaction between C_2H_4 and $H:^-$. As these two species approach each other, the pair of electrons of the $H:^-$ is donated to the C_2H_4. Identify the orbital on each of these species that is involved in this electron transfer. Identify the orbital in which the electron pair is found after the reaction is over.

2.7* Explain why an orbital with a node between the two nuclei is of higher energy than one without a node.

2.8 Without considering relative energies, explain why an electron is more likely to be found in the combination orbital that forms from constructive interference?

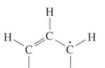

2.9 The radical ⟨structure⟩ is a resonance hybrid with two equivalent contributing structures. It is a relatively stable radical. Use a molecular orbital description to explain its relative stability.

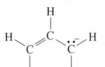

2.10 The anion ⟨structure⟩ is a resonance hybrid with two equivalent contributing structures. It is a relatively stable anion. Use a molecular orbital description to explain its relative stability.

2.11 The compound H $\underset{H}{\overset{H}{C}}=C=C\underset{H}{\overset{H}{}}$ is not conjugated. Explain.

2.12 Considering only the π electrons, list the possible electronic transitions that can occur in ethylene. Indicate which one is at longest wavelength and which one is at shortest wavelength. [Hint: See Table 1.3 for bond energies.]

2.13 Considering only the π electrons, list the possible electronic transitions that can occur in 1,3-butadiene. Indicate which one is at longest wavelength and which one is at shortest

*More challenging questions.

wavelength. (Assume all C—H bonds are the same strength.) [Hint: See Table 1.3 for bond energies.]

2.14* Considering only the π electrons and the lone pair, list the possible electronic transitions

that can occur in . Indicate which one is at longest wavelength and which one is at shortest wavelength. [Hint: See Table 1.3 for bond energies.]

2.15 The longest-wavelength radiation absorbed by 1,3-butadiene is 217 nm. Calculate the energy difference between π_2 and π_3^* from this measurement. ($c = 3.0 \times 10^8$ m/s, $h = 6.6 \times 10^{-34}$ J s, $N = 6.0 \times 10^{23}$ mol^{-1})

2.16 The longest-wavelength radiation absorbed by ethylene is 174 nm. Calculate the strength of the π bond from this measurement. ($c = 3.0 \times 10^8$ m/s, $h = 6.6 \times 10^{-34}$ J s, $N = 6.0 \times 10^{23}$ mol^{-1})

2.17 Draw schematic diagrams of the type in Figure 2.11 for the six π orbitals of a six-atom conjugated system, arranged in order of increasing energy.

Chapter 3
Acids and Bases

The tens of millions of organic compounds known undergo hundreds of millions of reactions. In order to understand and predict the course of these reactions in any manageable way, we need to systematize our knowledge and identify principles and patterns of chemical behavior. One important approach is to identify and define groups of compounds or chemical species that display similar chemical behavior. Two of the most important groups are acids and bases.

All compounds identified as acids undergo essentially the same chemical reactions when they act as acids. All compounds identified as bases undergo essentially the same chemical reactions when they act as bases. And most important, any acid will react with any base in the same way. The usefulness and scope of this approach depends on how many compounds we can define as acids and as bases.

3.1 | Brønsted Acids and Bases

The first reasonably general definition of acids and bases was proposed by J. N. Brønsted in the first quarter of the twentieth century.

1. An acid (also called a Brønsted acid) is a chemical species that can donate a proton (H^+).
2. A base (also called a Brønsted base) is a chemical species that can accept a proton.

These definitions make it very clear why an acid and a base react with each other. They also make it very clear what that reaction will be. The reaction between a Brønsted acid and a Brønsted base is the transfer of a proton from the acid to the base. Such reactions are almost always reversible and almost always occur in solution.

These definitions seem very simple and general. But they have some important ramifications that help us understand many important aspects of the chemical behavior of these compounds. We can represent a Brønsted acid by the symbol HA, which is not intended to convey any information about the charge of the acid. Acids can be neutral, positively charged, or negatively charged. But no matter what their charge, they must have an H, which can be donated by the acid as H^+. We can then symbolize this behavior with a skeleton equation that is *not* balanced with respect to electrical charge:

$$HA \rightleftharpoons A + H^+$$

The equation indicates that the acid HA is donating a proton. It does not indicate the recipient of this donation, and therefore it does not describe an actual chemical process.

An acid must donate a proton to something; free protons cannot exist to any significant extent under ordinary chemical conditions. The actual chemical process is a proton transfer from the acid to another chemical species that can accept the proton. The chemical species that can accept a proton is by definition a Brønsted base. We can represent a Brønsted base by the symbol B, which is not intended to convey any information about the charge of the base. Bases can also be neutral, positively charged, or negatively charged. We can symbolize this behavior with a skeleton equation that is *not* balanced with respect to electrical charge:

$$H^+ + B \rightleftharpoons BH$$

The equation indicates that the base B is the recipient of the proton. It does not indicate the donor of the proton, and therefore it does not describe an actual chemical process.

Before moving to an actual chemical equation, let us consider what information is conveyed by these two very simple skeleton equations. A key feature of acid–base reactions is that they are reversible, as is shown by the equilibrium sign in the equations. The donation of a proton by HA forms a species A that differs from HA only by having one proton fewer than HA. That means that if A were to accept a proton, it would form the chemical species HA; and therefore A must be a base. Clearly there is a close relationship between HA and A. Together they are called a conjugate acid–base pair. Remove a proton from HA, and we obtain A, which is called the conjugate base of the acid HA. Add a proton to A, and we obtain HA, which is called the conjugate acid of the base A.

Notice that the two skeleton equations we have written become identical if we change A to B or vice versa. We have simply switched left and right. So we can apply the same analysis to the second equation. BH is the conjugate acid of the base B, and B is the conjugate base of the acid BH.

These definitions allow us to predict the kinds of chemical species that will be Brønsted acids and Brønsted bases and to write specific equations for their behavior as acids and bases.

Most obviously, a Brønsted acid must have an H. If a species does not have an H, it cannot be a Brønsted acid. But it must also be an H that can be donated as a proton. In theory any H can be donated as a proton if the conditions are extreme enough, but the acids that we will generally encounter will be those that can donate a proton fairly easily. There are a number of factors that influence the ease with which a species can donate a proton. One factor is the polarity of the bond between the H that is to leave as a proton and the atom to which it is bonded. Since the electron pair of the bond remains behind when the proton leaves, the more electronegative the atom to which it is bonded, the easier it is for the proton to leave. For example, many acids have an O—H bond:

$$X—O—H \rightarrow X—O^- + H^+$$

The O—H bond is polarized with partial positive charge on H and partial negative charge on O. When it breaks, the O acquires a negative charge, which it can support because of its high electronegativity. The same reasoning explains why the hydrogen halides—HF, HCl, HBr, and HI—are acids.

Another factor that can facilitate proton donation is stabilization of the resulting conjugate base, usually by resonance. For example, a carboxylic acid such as formic acid readily donates a proton because its conjugate base, the anion which forms, is resonance stabilized (Figure 3.1).

Brønsted acids can be neutral compounds. Many familiar compounds such as H_2O, HF, H_2S, HCl, HCN, HNO_3, H_2SO_4, H_3PO_4, and $HClO_4$ are examples. Brønsted acids can also be cations. For example, NH_4^+, H_3O^+, PH_4^+, $CH_3NH_3^+$, and $C_2H_5^+$ are all proton donors. Brønsted acids can also be anions, often derived from polyprotic acids. Well-known examples are HCO_3^-, HSO_4^-, $H_2PO_4^-$, and HPO_4^{2-}. Given the identity of the acid, we can identify its conjugate base. It is the

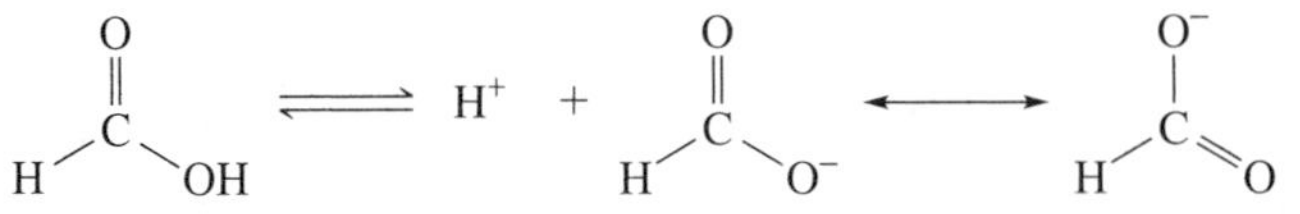

Figure 3.1

Resonance in the conjugate base of a carboxylic acid

result of the loss of an H and a decrease of the electrical charge by 1 (because of the loss of the positive proton). Some examples follow:

$$HF \rightleftharpoons H^+ + F^-$$

$$HCN \rightleftharpoons H^+ + CN^-$$

$$H_3PO_4 \rightleftharpoons H^+ + H_2PO_4^-$$

$$CH_3NH_3^+ \rightleftharpoons H^+ + CH_3NH_2$$

$$HCO_3^- \rightleftharpoons H^+ + CO_3^{2-}$$

Although they do not correspond to real chemical processes, these equations are still useful because they identify an acid and its conjugate base. Notice also that writing these equations in reverse does not change anything (because they are reversible) except that now we identify a base and its conjugate acid.

In some ways, the definition of a Brønsted base is more general than that of a Brønsted acid. While a Brønsted acid must have an H that it can donate as a proton, all a Brønsted base needs is a pair of electrons that it can use to form a bond to a proton. Most commonly the electrons will be a lone pair. All five of the bases shown above use a lone pair to form a bond to the proton. But in many cases, electrons in π bonds can also be used to form a bond to a proton.

Brønsted bases can be neutral compounds with lone pairs, such as compounds of N in its −3 oxidation state (ammonia, amines, etc.) or of O. The halogens also have lone pairs. But they almost never act as Brønsted bases when part of neutral compounds. Halogens in the −1 oxidation state rarely form more than one bond, and they are therefore not considered to be proton acceptors. Brønsted bases can also be compounds with multiple bonds, such as C_2H_4 and C_2H_2. In general, an anion has at least one lone pair, and therefore many Brønsted bases are anions. There are a few examples of Brønsted bases that are cations, but we shall not consider them.

Given the identity of the base, we can identify its conjugate acid. It is the result of the gain of an H and an increase of the electrical charge by 1 (because of the gain of the positive proton). Some examples of bases with lone pairs can be seen above. Simply reverse any of the five example reactions that identify acids. A few additional examples follow:

$$HS^- + H^+ \rightleftharpoons H_2S$$

$$CH_3CO_2^- \text{ (acetate ion) } + H^+ \rightleftharpoons CH_3CO_2H \text{ (acetic acid)}$$

$$NH_3 + H^+ \rightleftharpoons NH_4^+$$

$$HCO_3^- + H^+ \rightleftharpoons H_2CO_3$$

Notice that the last equation in each set of examples has bicarbonate ion as a reactant. It can act as an acid, donating a proton and forming its conjugate base, carbonate ion. It can act as a base, accepting a proton and forming its conjugate acid, carbonic acid. Substances that can act as either an acid or a base are said to be **amphoteric**.

An example of a base using a pair of π electrons to accept the proton is

In Section 1.3, we showed, using curved arrows, the way in which electron bookkeeping is done in organic chemistry. This kind of electron bookkeeping works very well for showing acid–base reactions in which bonds are broken and made. The tail of the arrow is on the electron pair of the base that is being used to bond the proton that it is accepting. The head of the arrow points to the proton, which is going to share the electron pair and bond to the base.

One of the advantages of the Brønsted theory is that it ties together the behavior of a great many chemical species in a simple framework.

3.2 | Acid–Base Reactions

The reaction between an acid and a base is one of the most fundamental in chemistry. In the Brønsted formulation, this reaction is a proton transfer from an acid to a base. As we have seen, when an acid donates a proton it is transformed into its conjugate base, and when a base accepts a proton it is transformed into its conjugate acid. Therefore the reaction between an acid and a base will include two conjugate acid–base pairs. Using the notation of our skeleton equations, which do not show electrical charge balance, the reaction of an acid and a base is

$$HA + B \rightleftharpoons A + BH$$

One conjugate acid–base pair is HA and A; the other is B and BH. Using this framework, we can write the chemical equation for the reaction of any Brønsted acid with any Brønsted base. Here are some examples that illustrate a variety of types of acids and bases:

1. Neutral acid + neutral base

$$CH_3CO_2H \text{ (acetic acid)} + CH_3NH_2 \rightleftharpoons CH_3CO_2^- + CH_3NH_3^+$$

    acid   base   conj base   conj acid

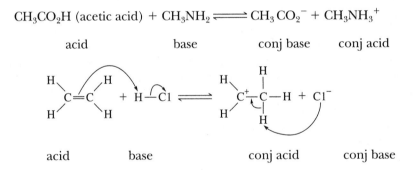

    acid   base   conj acid   conj base

2. Neutral acid + negative base

$$HBr + CN^- \rightleftharpoons Br^- + HCN$$

   acid   base   conj base   conj acid

3. Positive acid + neutral base

$$C_2H_5^+ + H_2O \rightleftharpoons C_2H_4 + H_3O^+$$

   acid   base   conj base   conj acid

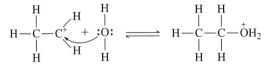

4. Positive acid + negative base

$$NH_4^+ + CO_3^{2-} \rightleftharpoons NH_3 + HCO_3^-$$

   acid   base   conj base   conj acid

5. Negative acid + neutral base

$$HPO_4^{2-} + NH_3 \rightleftharpoons PO_4^{3-} + NH_4^+$$

$$\text{acid} \qquad \text{base} \qquad \text{conj base} \quad \text{conj acid}$$

6. Negative acid + negative base

$$HSO_4^- + OH^- \rightleftharpoons SO_4^{2-} + H_2O$$

$$\text{acid} \qquad \text{base} \qquad \text{conj base} \quad \text{conj acid}$$

Given a compound that can be identified as an acid and given a compound that can be identified as a base, we can write the products of the chemical reaction between them. We are thus able to put thousands of different reactions into a systematic framework.

3.3 | Acid and Base Strength

While being able to predict the products that form when two reactants are mixed is very useful, that is only part of the story. We also need to be able to predict the extent to which the reaction proceeds. We approach this question by defining a property of acids and bases called their **strength.** We could define the strength of an acid by reference to the skeleton electrically unbalanced equation we used to define an acid: $HA \rightleftharpoons A + H^+$ and say that the greater is the equilibrium constant, K, for this skeleton reaction, the stronger is the acid HA. In other words, the stronger the acid the more readily it gives up its proton. While this definition is clear, it is not practical because we cannot measure an equilibrium constant of a reaction that does not take place. Recall that an acid must donate a proton to something.

Nevertheless, this approach gives us some important information about conjugate acid–base pairs. The relative strength of HA and the relative strength of its conjugate base A are defined by the relative amount of each when the system reaches equilibrium. The stronger is HA, the larger is K; the further to the right is the position of equilibrium, the greater is the relative amount of its conjugate base at equilibrium. The relatively large amount of the conjugate base present at equilibrium means that the conjugate base does not readily accept the proton. In other words, it is relatively weak. This direct relationship between the strengths of the two members of a conjugate acid–base pair is at the heart of the Brønsted theory. The greater the strength of one member of a conjugate acid–base pair, the weaker is the other one. For example, HCl is a stronger acid than HF, so Cl^- is a weaker base than F^-.

In order to put the relative strengths of Brønsted acids or the relative strengths of Brønsted bases on a quantitative footing, we need to use a real reaction whose equilibrium constant we can measure. The reaction will be the transfer of a proton from the acid to a base. Using the notation of our skeleton equations, which do not show electrical charge balance, the reaction can be written as

$$HA + B \rightleftharpoons A + BH$$

The equilibrium constant, K, of this reaction depends on the strengths of both HA, the acid, and B, the base. But if we look at this reaction for a series of acids reacting with one specific base, the value of the equilibrium constant for the reaction of each acid tells us about its strength relative to the other acids.

The most convenient base to use for this purpose is H_2O. The details of the equilibrium reaction depend on whether the acid is neutral, positive, or negative. The equilibrium constant for this reaction is called K_a (or K_1, K_2, or K_3 for polyprotic acids). Here is an example of each type of acid:

$$HCN + H_2O \rightleftharpoons CN^- + H_3O^+$$

$$NH_4^+ + H_2O \rightleftharpoons NH_3 + H_3O^+$$

$$HPO_4^{2-} + H_2O \rightleftharpoons PO_4^{3-} + H_3O^+$$

Notice that the base (H_2O) and its conjugate acid (H_3O^+) are the same in all three reactions. We could use a similar strategy to determine the relative strengths of a series of bases. But it is not really necessary. The data on the relative strengths of acids provide that information as well, because of the connection between the strengths of both members of a conjugate acid–base pair. Remember the stronger the acid the weaker its conjugate base, and vice versa.

For example, since we know that ammonium ion, NH_4^+, is a stronger acid than methyl ammonium ion, $CH_3NH_3^+$, we also know that ammonia, NH_3, is a weaker base than methyl amine, CH_3NH_2. Or since we know that HNO_3 is a stronger acid than HNO_2, we also know that its conjugate base NO_3^- is a weaker base than NO_2^-, the conjugate base of HNO_2. So while an equilibrium constant K_b can also be defined for a series of bases reacting with a specific acid (usually water), it is not generally done in organic chemistry. Some of these relationships, based on the proton transfer from the acid to H_2O acting as a base, are shown qualitatively in Table 3.1 for some common acids and bases.

A list of acids in order of increasing strength such as the one in Table 3.1 can be constructed only if appropriate measurements have been made. But even without the actual numerical data, such a list can be very useful. We can write many acid–base reactions of the following form:

$$\text{Acid} + \text{base} \rightleftharpoons \text{conj acid} + \text{conj base}$$

Those that are most useful will be the ones that proceed substantially toward completion. In other words, an acid–base reaction is useful if it has a relatively large K. The reaction will have a relatively large K when the following conditions exist. The starting acid on the left is stronger than the product acid on the right, and the starting base on the left is stronger than the product base on the right. We can determine directly from Table 3.1 if these conditions exist. The reaction between an acid and any base above it in the list will be favorable and will have a relatively large K.

Table 3.1	Relative Strengths of Acids and Their Conjugate Bases	
Increasing strength Acid	**Conjugate base**	**Increasing strength**
H_2O	OH^-	
HPO_4^{2-}	PO_4^{3-}	
H_2O_2	HO_2^-	
$CH_3NH_3^+$ (methyl ammonium)	CH_3NH_2	
$HClO$	ClO^-	
$HCOOH$ (formic acid)	$HCOO^-$	
HSO_4^-	SO_4^{2-}	
H_3O^+	H_2O	
HNO_3	NO_3^-	

Consider the acid HClO, which is in the middle of the list. Its reaction with any base above it—OH^-, PO_4^{3-}, HO_2^-, or CH_3NH_2—will be favorable and will have a relatively large K. The reason is that the conjugate acids of all the bases above it—H_2O, HPO_4^{2-}, H_2O_2, and $CH_3NH_3^+$—are all weaker acids than HClO, as the table shows. Its reaction with any base below it—$HCOO^-$, SO_4^{2-}, H_2O, and NO_3^-—will be unfavorable. The reason is that the conjugate acids of all the bases below it—HCOOH, HSO_4^{2-}, H_3O^+, and HNO_3—are stronger acids than HClO. Even this small table with only 9 acids gives us useful information about more than 70 different reactions.

The table actually gives information about many additional organic acids and bases. The members of a given class of organic acids tend to have similar strengths. As a result, many other members of a class of compounds called carboxylic acids—such as acetic acid, butyric acid, and lactic acid—are similar in acid strength to formic acid, which is in the table. In the same way, methyl ammonium ion is a member of a very large class of compounds, the protonated amines. Many other protonated amines such as those from dimethyl amine, ethyl amine, diethyl amine, and ephedrine will all have acid strengths similar to methyl ammonium ion. The unprotonated amines themselves will have similar base strengths to methyl amine.

In general chemistry, we do many calculations using these equilibrium constants. In organic chemistry, it is much more important to know the *relative* strengths of different acids and bases than to assign actual numbers. In part, this difference reflects the fact that most organic reactions do not take place in water, but in a range of different solvents. It would be impractical and usually not necessary to measure a series of K_a values in many different solvents. But we would also like to know more than simply that one acid is stronger than another. We would like to know at least roughly by how much. We use measurements made with reference to water acting as a base for this purpose.

Defining acid and base strengths using measured equilibrium constants has certain practical limitations. We are not able to measure directly or accurately very large or very small equilibrium constants. The value of the equilibrium constant depends on the base we choose to accept a proton from a series of acids. For example, when we choose H_2O as the base, some acids are so strong that the proton transfer reaction will proceed essentially to completion. In such cases, we cannot measure the equilibrium constant, and therefore we cannot determine the relative strengths of these acids in water. We say that they are strong, and we use other methods for determining their relative strengths.

Among the familiar strong acids are HCl, HBr, HI, H_2SO_4, HNO_3, and $HClO_4$. In water these acids all seem equally strong because water is a sufficiently strong base to drive the proton-transfer reaction to completion. This effect, which is the result of the base strength of water, is called the **leveling effect**. We can, however, determine the relative strengths of these acids, for example, by using a weaker base than water, one that does not drive the proton-transfer reaction to completion. We then extrapolate back to water.

There are similar problems in trying to determine the relative strengths of very weak acids. In such cases, water is not a strong enough base to bring about any measureable extent of proton transfer. So the relative strength of weak acids such as NH_3 (conjugate base NH_2^-) or CH_4 (conjugate base CH_3^-) cannot be measured in water. A stronger base is needed to bring about a measureable proton transfer, and again we extrapolate back to water.

An enormous number of equilibrium constants of proton transfers from acids to water have been measured, directly or indirectly, and tabulated. These equilibrium constants are designated by the symbol K_a and are called acid dissociation constants (an inaccurate name). These acid dissociation constants range from very large to very small numbers. It is convenient to express a very large or very small number as its negative logarithm (to the base 10), which is easier to write than an exponential. The symbol p is used to designate the negative logarithm. The relationship is

$$-\log K_a = pK_a$$

Data about the relative strengths of acids (and indirectly of their conjugate bases) are often presented as pK_a values. In general chemistry K_a (or K_1) is defined only for neutral acids. Data about the acid strength of positively charged acids are given indirectly as the K_b of their conjugate bases. Data about anionic acids, derived from polyprotic acids, are given as K_2 or K_3. But in organic chemistry we shall use K_a and pK_a to describe the strength of all acids, neutral or charged. Keep in mind that the lower the value of pK_a, the stronger the acid. Strong acids have negative pK_a values, and very weak acids have large positive pK_a values.

Most acids, for which an equilibrium constant is actually measureable in water, are called weak acids. We can use these equilibrium constants (often presented as pK_a values) to formulate a reasonably accurate quantitative picture of the relative strengths of such acids and their conjugate bases. There are also two other categories of acids called strong acids and very weak acids. Many experiments using suitable bases to replace water or more indirect methods have provided data on the relative strengths of these types of acids and their conjugate bases.

An understanding of the relative strengths of acids and the relative strengths of bases is a powerful tool in understanding their chemical behavior. We know that once we identify a compound as an acid and another compound as a base, we can write the products of the chemical reaction between them. But we cannot predict with any accuracy the extent to which the reaction proceeds without also having information about their relative strengths. Clearly the reaction between a stronger acid and a stronger base will be more favorable and will proceed further toward completion than the reaction between a weaker acid and a weaker base. But we would also like to know by how much.

In addition, many of the chemical systems we will encounter will contain more than one acid and/or more than one base. In order to predict the chemical behavior of such systems, we rely on the relative strengths of the acids and the bases in the system. The principle is straightforward. The most important and most favorable reaction in such a system is the one between the strongest acid and the strongest base. We can also determine whether there are other acid–base reactions that will compete. Not only can we predict chemical behavior based on such information but we can modify the way in which we carry out reactions to favor the formation of the products we want if they are not those from the strongest acid reacting with the strongest base. Table 3.2 lists pK_a values for a number of familiar inorganic and organic acids in order of increasing acid strength.

Table 3.2	Relative Strengths of Acids and Their Conjugate Bases			

pK_a	Acid		Conj base	
50	CH_4	Methane	CH_3^-	Methyl carbanion
35	NH_3	Ammonia	NH_2^-	Amide
25	C_2H_2	Acetylene	C_2H^-	Acetylide
16	C_2H_5OH	Ethanol	$C_2H_5O^-$	Ethoxide
15.75	H_2O	Water	OH^-	Hydroxide
15.2	CH_3OH	Methanol	CH_3O^-	Methoxide
12.33	HPO_4^{2-}	Hydrogen phosphate	PO_4^{3-}	Phosphate
11.62	H_2O_2	Hydrogen peroxide	HO_2^-	Hydroperoxide
10.81	$C_2H_5NH_3^+$	Ethyl ammonium	$C_2H_5NH_2$	Ethyl amine
10.66	$CH_3NH_3^+$	Methyl ammonium	CH_3NH_2	Methyl amine
10.64	HIO	Hypoiodous acid	IO^-	Hypoiodite
9.99	C_6H_6O	Phenol	$C_6H_5O^-$	Phenoxide
9.81	$(CH_3)_3NH^+$	Trimethyl ammonium	$(CH_3)_3N$	Trimethyl amine
9.31	HCN	Hydrocyanic acid	CN^-	Cyanide

pKa	Acid	Name	Conjugate base	Name
9.25	NH_4^+	Ammonium	NH_3	Ammonia
8.68	$HBrO$	Hypobromous acid	BrO^-	Hypobromite
7.50	$HClO$	Hypochlorous acid	ClO^-	Hypochlorite
7.21	$H_2PO_4^-$	Dihydrogen phosphate	HPO_4^{2-}	Hydrogen phosphate
6.37	H_2CO_3	Carbonic acid	HCO_3^-	Bicarbonate
5.21	$C_5H_5NH^+$	Pyridinium	C_5H_5N	Pyridine
4.76	CH_3CO_2H	Acetic acid	$CH_3CO_2^-$	Acetate
4.63	$C_6H_5NH_3^+$	Anilinium	$C_6H_5NH_2$	Aniline
3.73	HCO_2H	Formic acid	HCO_2^-	Formate
3.45	HF	Hydrogen fluoride	F^-	Fluoride
2.85	$ClCH_2CO_2H$	Chloroacetic acid	$ClCH_2CO_2^-$	Chloroacetate
2.12	H_3PO_4	Phosphoric acid	$H_2PO_4^-$	Dihydrogen phosphate
1.99	HSO_4^-	Hydrogen sulfate	SO_4^{2-}	Sulfate
1.96	$HClO_2$	Chlorous acid	ClO_2^-	Chlorite
1.48	Cl_2CHCO_2H	Dichloroacetic acid	$Cl_2CHCO_2^-$	Dichloroacetate
0.7	Cl_3CCO_2H	Trichloroacetic acid	$Cl_3CCO_2^-$	Trichloroacetate
0.3	F_3CCO_2H	Trifluoroacetic acid	$F_3CCO_2^-$	Trifluoroacetate
−1	$HClO_3$	Chloric acid	ClO_3^-	Chlorate
−1.76	H_3O^+	Hydronium	H_2O	Water
−2	HNO_3	Nitric acid	NO_3^-	Nitrate
−2	CH_3SO_3H	Methane sulfonic acid	$CH_3SO_3^-$	Methane sulfonate
−3	H_2SO_4	Sulfuric acid	HSO_4^-	Hydrogen sulfate
−7	HCl	Hydrochloric acid	Cl^-	Chloride
−9	HBr	Hydrogen bromide	Br^-	Bromide
−10	HI	Hydriodic acid	I^-	Iodide
−10	$HClO_4$	Perchloric acid	ClO_4^-	Perchlorate

As we have already seen, we can use the relative positions of conjugate acid–base pairs in this table to predict whether a given acid–base reaction is favorable or unfavorable. Remember, the reaction between an acid and any base above it in the table will be favorable and will have a relatively large K, and the reaction between an acid and any base below it in the table will be unfavorable and have a relatively small K.

Using the pK_a data in Table 3.2, we can predict just how favorable or unfavorable the reaction will be. We can calculate the pK (and thus the K) of the reaction of any acid in the table with any base in the table. The relationship is

$$pK = pK_a \text{ of the acid on the left} - pK_a \text{ of the acid on the right}$$

Notice that when the acid on the right is closer to the top of the table, it has a larger pK_a than the acid on the left. When we subtract its pK_a from that of the acid on the left, the result is a negative number. When the pK is negative, K is greater than 1 and the reaction is favorable. When the acid on the left is closer to the top of the table, the result of the subtraction is a positive number. A positive pK means that K is less than 1.

Example 3.1

Find the pK and K for the reactions that take place when each of the following compounds are mixed.

(a) hydrochloric acid + methyl amine

(b) acetylene + hydroxide

(c) hydrocyanic acid + phenoxide

Solution

(a) *Step 1:* The first and most important step in solving any problem about a chemical reaction is to **write the chemical reaction.**

$$HCl + CH_3NH_2 \rightleftharpoons CH_3NH_3^+ + Cl^-$$

Step 2: Identify the acid on the left; it is HCl. Identify the acid on the right; it is $CH_3NH_3^+$.

Step 3: Find their pK_a values in Table 3.2. The values are -7 and 10.66, and subtract the one for the product acid—in this case, methyl ammonium from the reactant acid, HCl.

$-7 - 10.66 = -17.66 = pK$ of the reaction (Remember p means the negative log)

$$K = 10^{17.66} = 4.6 \times 10^{17}$$

As expected from the position of the starting acid, which is much lower than the starting base in the table, the reaction is very favorable.

(b) *Step 1:*

$$C_2H_2 + OH^- \rightleftharpoons H_2O + C_2H^-$$

Step 2: The acid on the left is C_2H_2, and the acid on the right is H_2O.

Step 3: The respective pK_a values are 25 and 15.75.

The pK of the reaction = $25 - 15.75 = 9$

$$K = 10^{-9}$$

As expected from the relative position of the starting acid, which is well above the starting base in the table, the reaction is very unfavorable.

(c) *Step 1:*

$$HCN + C_6H_5O^- \rightleftharpoons C_6H_5OH + CN^-$$

Step 2: The acid on the left is HCN, and the acid on the right is C_6H_5OH.

Step 3: The respective pK_a values are 9.31 and 9.99.

The pK of the reaction + $9.31 - 9.99 = -0.68$

$$K = 10^{0.68} = 4.8$$

Even though the acid and base are relatively close in the table, the reaction is still reasonably favorable because the base is above the acid.

3.4 | Acid–Base Strength and Structure

The relative strength of an acid depends on the relative stability of the acid and its conjugate base, as we have seen. The more stable an acid is relative to its conjugate base, the weaker is the acid and the stronger is the conjugate base. The reverse must also be true. The more stable the base, the weaker it is and the stronger is its conjugate acid. Their relative stabilities are related to their structures. In many cases, there are a number of different competing factors, and we are often in the position of explaining, rather than predicting, the relative strength of an acid or a base.

Let us start by looking at the relative strengths of the simplest acids, the binary hydrides, which are compounds of two elements, one of which is H. There are two clear trends in the acid strength of these compounds. Their acid strength increases going down a column of the periodic table and increases going from left to right across a row of the periodic table.

Consider the series of acids HX, where X is the usual symbol for halogen. Experimentally we find that HF is the weakest of these acids and acid strength increases going down the column, so that HI is the strongest. We can explain this trend by the decrease in the strength of the hydrogen–halogen bond. The strongest bond is the HF bond and the weakest is the HI bond. Their pK_a values (extrapolated to water for strong acids) are

Acid	HF	HCl	HBr	HI
pK_a	3.45	−7(approx)	−9(approx)	−10(approx)

Now consider the hydrides of the nonmetals in the second row of the periodic table: CH_4, NH_3, H_2O, and HF. Experimentally we find that CH_4 is the weakest of these and acid strength increases, so that HF is the strongest. This trend is the reverse of what we would predict on the basis of bond strength. So we explain the results by the increase in electronegativity as we go from carbon to fluorine. As we have mentioned, as electronegativity increases, the electron pair of the H—A bond is held more closely by A and so dissociation to H^+ and A^- becomes easier. Their pK_a values (extrapolated to water for very weak acids) are

Acid	CH_4	NH_3	H_2O	HF
pK_a	50 (approx)	35(approx)	15.75	3.45

Apparently there are two main factors working in opposite directions that determine the relative acid strength in these binary hydrides. In the hydrides of the halogens, bond strength wins. In the hydrides of the second-row nonmetals, electronegativity wins. Without the experimental data, it would be extremely difficult to predict the observed trends. We will encounter many situations in organic chemistry where there are at least two factors working in opposite directions. It will generally be very difficult to make predictions in these situations, but it will be relatively easy to explain observed trends on the basis of which factor is more important for the situation.

Many common acids are **oxyacids**, so named because they contain hydrogen, oxygen, and a third element that is usually a nonmetal. The acidic hydrogens are always bonded to oxygen in an oxyacid. Organic oxyacids (where the nonmetal is C) generally have additional hydrogens that do not affect acid strength in any important way. Some typical oxyacids are

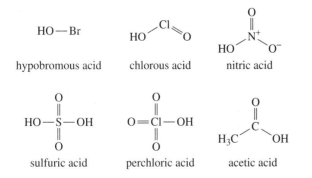

hypobromous acid   chlorous acid   nitric acid

sulfuric acid   perchloric acid   acetic acid

The strength of an oxyacid can be related to the polarity of the O—H bond that breaks to donate the proton. If the nonmetal is very electronegative, electrons are drawn further away from the H and the oxyacid is stronger. In the series HClO,

HBrO, and HIO, there is a decrease in acid strength that parallels the decrease in electronegativity from Cl to I. Their pK_a values are

Acid	HClO	HBrO	HIO
pK_a	7.50	8.68	10.64

Many of the strong oxyacids have more electronegative nonmetals. Examples are HNO_3 and H_2SO_4. Many of the weaker oxyacids have less electronegative nonmetals. Examples are H_3PO_4 and H_2CO_3. Their pK_a values are

Acid	H_2SO_4	HNO_3	H_3PO_4	H_2CO_3
pK_a	−3 (approx)	−2 (approx)	2.12	6.37

The polarity of the O—H bond is also influenced by the number of oxygens in the oxyacid, or, stated more generally, by the oxidation number of the nonmetal. In general, the higher the oxidation number, the stronger the acid. A good example of this effect can be seen in the oxyacids of Cl. There is a regular and dramatic increase in acid strength in the series HClO, $HClO_2$, $HClO_3$, and $HClO_4$. The last, perchloric acid is one of the strongest acids known. The pK_a values of the oxyacids of Cl (extrapolated to water for strong acids) are

Acid	HClO	$HClO_2$	$HClO_3$	$HClO_4$
pK_a	7.50	1.96	−1.0	−10 (approx)

The general formula of an oxyacid can be written as H_mYO_n, where Y is the nonmetal. The subscript m indicates the number of acidic hydrogens (usually bonded to O), and the subscript n indicates the number of oxygens. Remember that oxyacids can sometimes have hydrogens that are not acidic. For example, 3 of the 4 hydrogens in acetic acid are bonded to carbon and are therefore so weakly acidic that they do not count. There is a rule based on this general formula that incorporates these effects. The value of $n - m$, where n and m are the subscripts in the general formula for oxyacids, gives information about the relative strengths of many oxyacids. Generally a large value of $n - m$ indicates that the oxyacid is strong. For $HClO_4$, $n - m$ is $4 - 1 = 3$, and $HClO_4$ is a very strong acid. For H_2SO_4, $n - m$ is $4 - 2 = 2$, and H_2SO_4 is a strong acid. For H_3PO_4, $n - m = 1$, and H_3PO_4 is a weak acid. For HClO, $n - m$ is 0, and HClO is an even weaker acid.

All of these effects can be summarized in one simple generalization that is especially useful for organic acids: *Electron withdrawal increases acid strength*. A clear cut example of this effect can be seen in a series of substituted acetic acids.

As the hydrogens bonded to the C of acetic acid are replaced by more electronegative elements such as halogens, the strength of the resulting acid (Figure 3.2) keeps increasing. In the series shown in the following table, iodoacetic acid is somewhat stronger than acetic acid itself, and the acidity keeps increasing. In water, trifluoroacetic acid is almost 100,000 times stronger than acetic acid. Some pK_a values are

Acid	Acetic	Iodo	Bromo	Chloro	Dichloro	Trichloro	Trifluoro
pK_a	4.76	3.12	2.90	2.85	1.48	0.7	0.5

The effect of electron withdrawal on acid strength can also manifest itself in other ways. The electronegativity of carbon itself varies depending on its hybridization. Electrons in s orbitals are closer to the nucleus than electrons in p orbitals. As a result the more s character there is in a hybrid orbital used for a C—H bond, the more electronegative is the carbon and the more polarized is the C—H bond. An sp hybrid is 50% s, while an sp^3 hybrid is 25% s. Therefore an H bonded to an

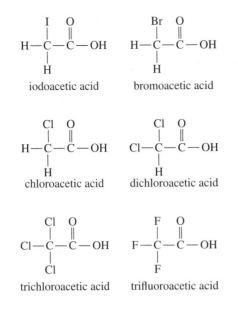

Figure 3.2
Halogenated Acetic Acids

sp-hybridized C is more acidic than one bonded to an sp^2-hybridized C, which is more acidic than an H bonded to an sp^3-hybridized C. We can see this effect in the series C_2H_6, C_2H_4, and C_2H_2. Their pK_a values are

Acid	C_2H_6	C_2H_4	C_2H_2
pK_a	50	44	25

Since electron withdrawal increases acid strength, we expect electron donation to decrease acid strength. It does. The clearest example of this effect is seen in the strengths of acids that can donate more than one proton, the polyprotic acids. Typical examples are H_2SO_4, H_3PO_4, H_2CO_3, and H_2S. In such acids, the K_a for the donation of the first proton (called K_1) is always greater than that for the donation of the second proton (called K_2), which is always greater than the one for the donation of the third proton. The reason for this trend is that removal of a proton creates negative charge. The negative charge results in more electron donation toward the bond of the acidic H. Some pK_a values are

Acid	H_3PO_4	$H_2PO_4^-$	HPO_4^{2-}	$C_2O_4H_2$ oxalic	$C_2O_4H^-$ oxalate	H_2CO_3	HCO_3^-
pK_a	2.12	7.21	12.33	1.27	4.27	6.37	10.25

For any trend in the strength of acids, we find the opposite trend for their conjugate bases. We have seen that acidity increases as we go down the halogen column of the periodic table from HF to HI. There is a corresponding decrease in base strength in their conjugate bases—the ions F^-, Cl^-, Br^-, and I^-—as we go down the column.

While in many situations, we will be able to predict the relative strengths of bases based on the strengths of their conjugate acids, it is sometimes easier to focus on the base itself. To accept a proton, a base must make available an electron pair. Therefore, we expect the strength of a base to be related to the availability of an electron pair. For example, $:NH_3$ is a relatively weak base whose strength is easily measured in water by measuring the strength of its conjugate acid, NH_4^+. But, $:NF_3$ is a very weak base, so weak that the strength of its conjugate acid is too great to measure in water. The lone pair in $:NH_3$ is available, but the electronegative F atoms in $:NF_3$ attract electrons so strongly that the lone pair is not available for bonding to a proton.

Based on this example and our analysis of acid strength, we can predict correctly that electron withdrawal decreases base strength by making the lone pair less available. It follows that electron donation increases base strength by making the electron pair more available. But just as in the case of acids, there can be other factors at work.

The availability of the lone pair does not explain why $:NH_3$ is a stronger base than $:PH_3$. We would predict the opposite because N is more electronegative than P, and so it should hold the lone pair more tightly. But we can explain this trend by considering their conjugate acids. As we have seen with the binary hydrides of the halogens, acidity increases going down the column. We see the same effect for the same reason in the conjugate acids of the binary hydrides in column VA. PH_4^+ is a stronger acid than NH_4^+ because the P—H bond is weaker than the N—H bond.

Notwithstanding this example, the generalization that electron donation increases base strength is a very useful one in organic chemistry. Virtually every base we will encounter will be a second-row element with a lone pair and most neutral bases will have the lone pair on N. So the effect of moving down a column on bond strength is not an issue.

An understanding of the relationship between structure and acid strength can be very useful. It will enable us to make predictions about the course of many acid–base reactions without reference to lists of pK_a values such as the one in Table 3.2. For example, the reactions of very strong acids with almost any base we are likely to encounter will be practical. Similarly the conjugate bases of very weak acids, bases such as CH_3^- or NH_2^-, will react with virtually any acid we are likely to encounter. On the other hand, very weak acids or the conjugate bases of very strong acids will generally not be very useful reagents for acid–base reactions.

3.5 | Lewis Acids and Bases

As useful as the Brønsted theory is, it still has some shortcomings. Since it defines an acid as a proton donor, it excludes from the category of acids compounds that have no protons to donate. It deals only with proton transfers. Fundamentally, a proton transfer is a reaction in which a bond is formed using a pair of electrons of one of the two substances (the Brønsted base) that bond to each other. Using that analysis G. N. Lewis proposed a very useful generalization of acid–base theory. He proposed that an acid is any chemical species that can accept an electron pair while forming a new covalent bond. We call such a species a Lewis acid. And he proposed that a base is any species that can donate an electron pair while forming a new covalent bond. We call such a species a Lewis base.

Although the two acid–base theories complement each other, there are some differences to keep in mind. In the Brønsted theory an acid is a proton donor. In the Lewis theory the proton is the acid. Brønsted acids are not themselves Lewis acids. For example, HCl, NH_4^+, or HCO_3^- clearly cannot accept an electron pair, but they can donate a proton, which can then accept an electron pair. The power of the Lewis theory is that it classifies many more chemical species as acids and allows us to predict and understand their chemical behavior. On the other hand, a Brønsted base and a Lewis base are defined in almost the same way. Either one is an electron pair donor. But the Brønsted base donates its electron pair only to a proton. A Lewis base can donate its electron pair to any electron acceptor. This formulation increases our understanding of a much broader range of chemical reactions.

Let us consider what kinds of species are Lewis acids. Our focus will be primarily on reactions of atoms that are in the second row of the periodic table. We can say unequivocally that to be a Lewis acid, the second-row atom must be able to accept the electron pair and form a bond to the Lewis base without expanding its octet. There are three different ways this bond formation can happen.

1. The atom can start with an incomplete octet.
2. The species can have at least one contributing resonance structure in which the atom has an incomplete octet.
3. Another bond to the atom can break while the new bond is forming.

Any species that satisfies one of these three conditions is a Lewis acid. Some simple examples in which the C has an incomplete octet are cations such as CH_3^+ and $C_2H_5^+$ in each of which the C atom that bears the plus charge has an incomplete octet. Such cations are extremely reactive and will combine very readily with any Lewis base. A typical reaction is

$$CH_3^+ + H_2O \longrightarrow CH_3OH_2^+$$

Any compound in which C is part of a multiple bond will have a contributing structure that puts positive charge on carbon. Examples are CH_2O, CH_2NH, and even C_2H_4. These compounds will react with a Lewis base to form a new σ bond to C, while the π bond of the multiple bond breaks. A typical reaction is the one between formaldehyde and cyanide ion, in which the negative Lewis base attaches to the relatively positive C.

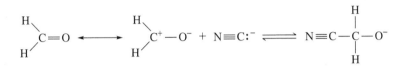

When a C is bonded to an electronegative atom by a single bond, the bond is polarized to make the C relatively positive. A Lewis base can form a bond to such a carbon if the bond to the electronegative atom simultaneously breaks. Examples of such compounds are CH_3Cl, CH_3Br, and CH_3I. A typical reaction in which a bond breaks as a bond forms is

$$CH_3I + OH^- \longrightarrow CH_3OH + I^-$$

The situation with nitrogen is quite different. Many nitrogen-containing species have a positive charge on the N because it has four bonds rather than three. The positive nitrogen does not have an incomplete octet and therefore cannot accept an electron pair. Typical examples are NH_4^+ and $CH_3NH_3^+$, both of which are Brønsted acids, but not Lewis acids. They cannot accept an electron pair, because they would have to expand their octet to do so. Similarly, oxygen-containing compounds that have a positive charge on O because the O has three bonds rather than two are not Lewis acids. They cannot accept an electron pair because their octets are already complete.

One of the great successes of the Lewis theory was its explanation of the chemistry of compounds of Group IIIA elements such as B and Al. These compounds have an incomplete octet and therefore can accept an electron pair. Typical examples are BF_3 and $AlCl_3$, which are strong Lewis acids. They react vigorously with compounds that have lone pairs such as NH_3 or H_2O in a reaction that is classified as an acid–base reaction in Lewis theory. These types of reactions generally proceed essentially to completion.

$$BF_3 + :NH_3 \longrightarrow F_3\overset{-}{B}-\overset{+}{N}H_3$$

Because both a Brønsted base and a Lewis base provide an electron pair to form a bond, all Brønsted bases are Lewis bases and vice versa. The difference between them is that for Brønsted bases we consider only the formation of a bond with a proton, while for Lewis bases we consider the formation of a bond with any species that can accept an electron pair. Brønsted bases always form a bond to the same thing, a proton, so we can compare their strengths directly. But when we consider the strength or behavior of Lewis bases, there can be a complication because Lewis bases can form bonds to many different elements. The details of the properties of Lewis bases often depend on the nature of the Lewis acid to which they are donating their electron pair.

By far the two most important types of acids in organic chemistry are proton donors or any species in which the electron pair acceptor is a carbon. For a proton donor, we use the Brønsted theory. As we have seen, we determine relative acid–base strengths by considering the equilibrium reaction in which the Brønsted base accepts the proton.

$$H^+ + B \rightleftharpoons BH$$

We have discussed in Section 3.4 the various structural factors that can influence Brønsted acid–base strength. Because of the small size of the proton, all of these factors relate to the electronic structure of the proton acceptor, the Brønsted base. Remember, however, that although we find it convenient to write protons as H^+, they do not exist as free protons.

When a Lewis base reacts with a Lewis acid in which the electron pair acceptor is a carbon, a number of other factors can come into play. The carbon is always part of a chemical species that is substantially larger than a proton. As a Lewis base approaches such a species, it will encounter other atoms that impede its path. The extent to which its path is impeded will depend on the size of the Lewis base and the size and locations of the other atoms in the Lewis acid.

The influences of size on chemical properties are called *steric effects*. They often play an important role in determining the properties of organic molecules and the course of organic reactions.

3.6 | Nucleophiles and Electrophiles

As we have seen, many reactions can be understood as the combination of a Lewis acid and a Lewis base. A common method of classifying chemical reagents is based on the Lewis acid and Lewis base definition. Any reagent that seeks an electron pair is called an **electrophile**, which is from the Greek for "lover of electrons." Electrophile is simply another name for Lewis acid. Any reagent that seeks substances that are electron deficient is called a **nucleophile**, which is from the Greek for "lover of nuclei." Nucleophile is simply another name for Lewis base. We tend to use these designations instead when either the Lewis acid or the Lewis base in the reaction involves carbon.

When an electrophile and a nucleophile find each other, a reaction can occur. A simple example of such a reaction is

$$C_2H_5^+ + Br^- \longrightarrow C_2H_5Br$$

The electrophile is the $C_2H_5^+$ cation, and the nucleophile is the Br^- anion. This type of reaction, which makes a bond without breaking any bonds, is very favorable. This type of process in which a bond is formed between two atoms, using an electron pair brought by one of them, is a common, fundamental chemical reaction. In many reactions the bond formation is just one part of a more complicated overall process.

A more subtle example of a reaction between a nucleophile and an electrophile is

$$C_2H_5I + Br^- \rightleftharpoons C_2H_5Br + I^-$$

Again the Br^- is the nucleophile. But the electrophile is the compound C_2H_5I, one of whose C atoms is relatively positive. Its C—I bond can break as the C—Br bond forms, thus avoiding octet expansion. Notice that in this reaction the products are also a nucleophile (I^-) and an electrophile C_2H_5Br.

A reaction in which a proton forms a bond to carbon is also a reaction between an electrophile and a nucleophile

$$C_2H_4 + H^+ \rightleftharpoons C_2H_5^+$$

The electrophile is the proton, and the nucleophile that provides the electron pair is the C_2H_4. The electron pair is in a π bond, which can easily break, so that the electron pair can be donated to the proton. This reaction is not as favorable as the one in which $C_2H_5^+$ is the electrophile because in addition to a bond being made, a bond is being broken.

In Figure 3.3, we track the movement of electron pairs using the curved arrow notation we introduced in Section 1.3 for these three examples.

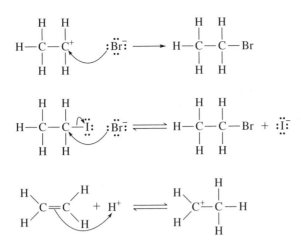

The first two examples show the kinds of differences that can be seen between a Brønsted base and a nucleophile. These reactions proceed significantly or almost entirely to completion because Br^- is a reasonably strong nucleophile, despite being an extremely weak Brønsted base.

3.7 | Nucleophilicity and Electrophilicity

The Brønsted theory precisely defines acid and base strength by measuring, directly or indirectly, the equilibrium constant of a reference reaction, the transfer of a proton from an acid to water, which is acting as a base. This approach works well because we can define a conjugate acid–base pair, where the species are related to each other by the gain or loss of a proton. Because of the much wider scope of the Lewis theory, there is no such simple consistent measurement we can use to define precisely the strengths of Lewis acids and bases.

Instead we define and measure properties called **nucleophilicity** and **electrophilicity**, as measures of the strengths of nucleophiles and electrophiles. These properties are usually measured by measuring rates of reactions. The reason we do not use equilibrium constants is because so many reactions between a nucleophile and an electrophile go to completion for all intents and purposes.

Measurements of nucleophilicity are generally easier to make than measures of electrophilicity and are easier to interpret. The general approach is similar to the one we used to measure the strength of Brønsted acids. We can choose a standard electrophile involving C and measure the rates of its reactions with a series of nucleophiles, at the same concentrations of nucleophile and electrophile. The faster the rate of the reaction, the greater the nucleophilicity. One widely used reaction for this purpose is the reaction of methyl iodide with a nucleophile. The product is one in which the nucleophile substitutes for the bound iodine, which is converted to I^-.

$$CH_3I + Nuc \longrightarrow CH_3Nuc + I^-$$

This general equation is not electrically balanced. The charge of the product, CH_3Nuc, will depend on the charge of Nuc. If the nucleophile has a charge of -1, the product will be neutral. If the nucleophile is neutral, the product will have a charge of $+1$. Some examples of these reactions are

$$CH_3I + Br^- \longrightarrow CH_3Br + I^-$$

$$CH_3I + OH^- \longrightarrow CH_3OH + I^-$$

$$CH_3I + H_2O \longrightarrow CH_3OH_2{}^+ + I^-$$

$$CH_3I + NH_3 \longrightarrow CH_3NH_3{}^+ + I^-$$

The most important factor that determines nucleophilicity is the basicity of the nucleophile, which is a measure of its ability to accept a proton (Brønsted) or to donate an electron pair (Lewis). But unlike Brønsted acidity and basicity, nucleophilicity can vary a great deal depending on the choice of the electrophile or the reaction conditions. There are a number of reasons for this variation. One important reason is size. The size of the nucleophile becomes more and more of a factor reducing nucleophilicity as the size of the electrophile increases. As the nucleophile gets close to the carbon to which it is seeking to bond, it bumps into other parts of the electrophile, slowing down the rate of the reaction. Another reason for variation is the solvent used. Many organic reactions, including ones between an electrophile and a nucleophile, can be carried out in a range of solvents. Interaction between a solvent and a nucleophile, what we call solvation, can affect its nucleophilicity. The stronger this interaction, the more difficult it is for the nucleophile to bond to the electrophilic carbon because the solvent gets in the way. There are a number of other effects on nucleophilicity that will be discussed in organic chemistry.

Exercises

3.1 Indicate whether each of the following chemical species is a Brønsted acid, a Brønsted base, neither, or both. Draw the conjugate base of the Brønsted acids and the conjugate acid of the Brønsted bases.

(a) HPO_4^{2-} (b) NCl_3 (c) CO_2 (d) $CHCl_3$ (e) CCl_4

(f) $CH_2{=}O$ (g) NH_3 (h) CH_3OH (i) BF_3

3.2 Some compounds that we normally think of as acids can also be Brønsted bases when they react with a stronger acid. Draw the structure of the conjugate acid of hypochlorous acid, HOCl.

3.3 Write equations for the reactions between the following pairs of species in aqueous solution. If no significant reaction takes place, write no reaction.

(a) bicarbonate ion + HI   (b) HF + cyanide ion

(c) HCl + PH_3 (d) ammonia + hydroxide ion

(e) acetic acid ($H_3C\overset{\overset{\displaystyle O}{\|}}{C}OH$) + methyl amine ($CH_3NH_2$)

(f) methoxide (CH_3O^-) + methyl ammonium ($CH_3NH_3^+$)

(g) ethoxide ($C_2H_5O^-$) + S^{2-} (h) $(CH_2OH)^+$ + NH_3 [Hint: see Exercise 3.1(f)]

3.4 For each of the reactions that you wrote in Exercise 3.3, identify the two conjugate acid–base pairs.

3.5 Although primarily thought of as an acid, formic acid, $H\overset{\overset{\displaystyle O}{\|}}{C}OH$, can also act as a Brønsted base. Draw the structures of two possible conjugate acids of formic acid.

3.6 Indicate which of the following species can be amphoteric and react with itself. Write the equation for the species reacting with itself and identify the two conjugate acid–base pairs in each reaction.

(a) HCO_3^- (b) NH_3 (c) H_3O^+ (d) H_3PO_4 (e) HSO_4^- (f) CH_3^+

3.7 Draw the structure of a species with the formula $C_2H_9N_2^+$ that is a Brønsted acid and a Brønsted base.

3.8 Consider an aqueous solution of equimolar concentrations of sodium phosphate (Na_3PO_4), sodium hypochlorite (NaClO), hydrogen peroxide (H_2O_2), and sodium bisulfate ($NaHSO_4$). Using the information in Table 3.1, write the reaction with the largest equilibrium constant.

3.9 Considering only the acids and bases listed in Table 3.1, write the acid–base reaction that has the largest K and the smallest K.

3.10 Of the following three acid–base reactions, predict which one has the largest K and which one the smallest K, using only the information in Table 3.1.

(a) Propionic acid and propyl amine; (b) acetic acid and ClO^-; (c) ethyl ammonium and acetate.

3.11 Using the data in Table 3.2, calculate the pK of the following reactions.

(a) Methane sulfonic acid and sodium sulfate

(b) Chloroacetic acid and sodium acetate

(c) Anilinium chloride and ammonia

(d) Trimethyl ammonium chloride and sodium formate

(e) Ethanol and sodium methoxide

(f) Methane and sodium amide

3.12 Arrange the following sets of acids in order of increasing pK_a (lowest pK_a first), without referring to any given pK_as.

(a) H_2O, H_2S, H_2Se, H_2Te

(b) SiH_4, PH_3, H_2S, HCl

(c) $HClO_4$, H_3PO_2, H_3PO_3, H_3PO_4, H_2SO_4

(d)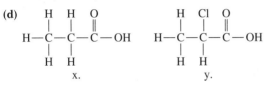

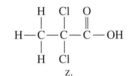

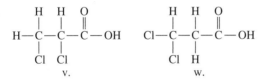

3.13 Arrange the following sets of bases in order of increasing base strength (weakest base first), without referring to any given pK_as.

(a) AsH_3, NH_3, PH_3, SbH_3

(b) x. CH_3NH_2.

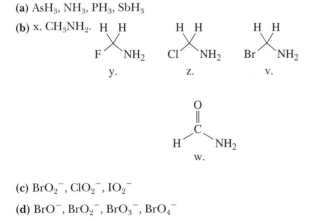

(c) BrO_2^-, ClO_2^-, IO_2^-

(d) BrO^-, BrO_2^-, BrO_3^-, BrO_4^-

3.14 Which of the following species are Lewis acids?

NH_4^+, CH_3^+, H_3O^+, HF, N_2O, NO^+, Cl^+, $Al(OH)_3$

3.15 Which of the following species are not Lewis bases?

CH_3^-, OH^-, H_2O_2, I^-, Cl^+, PH_4^+, CH_3OH, NH_2NH_2, BeF_2

3.16 Arrange the following Lewis acids in order of increasing strength (weakest one first) C_2H_4, CH_3N, CH_2O.

3.17 The reaction of C_2H_4 (ethylene) with HBr to form ethyl bromide, C_2H_5Br, can be thought of as occurring by two consecutive acid–base reactions. The first is a Brønsted acid–base reaction and the second is a Lewis acid–base reaction. Write these two reactions and identify the acid and base in each step.

3.18 The reaction between NaCN, HCl, and formaldehyde (CH_2=O) in aqueous solution is a two-step reaction. Both steps are acid–base reactions. Predict the final product by writing these two acid–base reactions.

3.19 Both H_2O and OH^- are nucleophiles. Predict which one is more nucleophilic and explain.

3.20 Although I^- is less basic than F^-, it is a stronger nucleophile in H_2O solution. It is, however, a weaker nucleophile in solvents that do not have OH or NH groups. Explain.

Chapter 4

Thermochemistry and Thermodynamics

4.1 | Heat

Almost every chemical or physical change is accompanied by the absorption or the evolution of heat. Thermochemistry is the branch of chemistry that studies heat effects and what they tell us about chemical structure and stability. Heat is a transfer of energy between a chemical system and its surroundings. When heat is absorbed by a chemical or physical change, the surroundings become cooler. We say the process is **endothermic.** When heat is evolved by a chemical or physical change, the surroundings become warmer. We say the process is **exothermic.** The overwhelming majority of chemical reactions that we shall study in organic chemistry are exothermic.

The quantity of heat evolved or absorbed during a chemical or physical change can give us valuable information about the structure and relative stability of all the components in the chemical system. This quantity of heat evolved or absorbed is often called the heat change or just the heat of the reaction. It can be measured using calorimetric techniques. The heats of an enormous number of chemical reactions have been measured and tabulated.

The actual value of the heat change accompanying a reaction depends not only on the specific reaction but on a number of other factors as well. The quantity of heat evolved or absorbed depends on the quantities of materials that are undergoing reaction. We all know that we will get more heat from burning two gallons of gasoline than from burning one gallon. So the heat is expressed using units that indicate the quantity of heat, usually kJ, and the quantity of material, usually mol. Heats of reactions are thus expressed in units of kJ/mol.

Another factor that affects the value of the heat change, but to a much lesser extent than quantity, is the conditions under which the reaction is carried out. Most organic reactions are carried out in containers that are open to the atmosphere, which means that they are taking place under conditions of essentially constant pressure. The heat change of a reaction carried out at constant pressure is called the enthalpy of the reaction, which is abbreviated ΔH. Whenever we refer to the heat of a reaction, we actually mean its ΔH.

The ΔH of an exothermic reaction is negative. For example, the ΔH for the combustion of natural gas

$$CH_4(g) + 2O_2(g) \longrightarrow CO_2(g) + 2H_2O(l)$$

is -890 kJ/mol.

The ΔH of an endothermic reaction is positive. For example, the ΔH for the dissociation of a chlorine molecule to two chlorine atoms

$$Cl_2(g) \longrightarrow 2Cl(g)$$

is 243 kJ/mol. Notice that the ΔH of the reverse reaction will always have the same numerical value, but the opposite sign as the forward reaction. This ΔH (243 kJ/mol) is also used to define the bond energy (really the bond enthalpy) of the Cl—Cl bond (Section 1.8).

The ΔHs of certain types of reactions are used very frequently. For example, it is relatively easy to measure the ΔHs of combustion reactions. The values of these ΔHs are used for a variety of purposes, including predicting the ΔHs of other reactions and the stabilities of certain compounds. Because they are used so frequently, it is convenient to define a quantity called the heat of combustion rather than talking about the heat of reaction of a combustion reaction. Unfortunately the heat of combustion is not defined exactly as the heat of reaction. For convenience, heats of combustion are expressed as positive values even though the reaction is exothermic. So $\Delta H_{comb} = -\Delta H_{react}$. This convention is used for some other reactions as well. For example, the addition of H_2 to certain organic compounds is an exothermic reaction called hydrogenation. But we define the heat of hydrogenation as a positive quantity.

4.2 | Bond Energy

Chemical change results from making and breaking chemical bonds. Heat is absorbed in order to break a chemical bond, and heat is evolved when a chemical bond is made. Clearly there will be a close relationship between the ΔH of the reaction and the bond energies of the bonds that are made and broken. But as we shall see, there may be a number of other factors that determine the relative stabilities of the reactants and products. Remember that bond energy is defined as the heat required to break a bond in the gas phase and that one electron of the pair of electrons of the bond is sent to each atom. For example, the H—H bond energy is defined by the reaction $H_2(g) \longrightarrow 2H(g)$.

We will start by considering a gas-phase reaction in which the only significant factors that determine stability are bond strengths.

$$CH_4 + Cl_2 \longrightarrow CH_3Cl + HCl$$

This reaction is easily observed to be quite exothermic, which means that the products are more stable than the reactants. For this reaction, it means that $CH_3Cl + HCl$ are more stable than $CH_4 + Cl_2$.

The reaction can be analyzed in terms of bond breaking and bond making. The CH_4 has 4 C—H bonds; the CH_3Cl has 3 C—H bonds. A C—H bond has been broken and a C—Cl bond has been made. The Cl_2 has a Cl—Cl bond, which has been broken, and the HCl has an H—Cl bond, which has been made. We can say that the total energy of an H—Cl bond and a C—Cl bond is stronger than the total energy of a C—H bond and a Cl—Cl bond. The Lewis structures show these changes.

Bond energy data (Tables 1.3 and 4.1) allow us to put this kind of analysis on a quantitative footing. Table 4.1 lists the bond energies of some common diatomic molecules in the gas phase.

Measurements have been made of the bond energies of many different C—H bonds and C—Cl bonds, and the average value of all these measurements provides a good working value called the average bond energy (Table 1.3) for each bond. For bonds in diatomic molecules such as Cl_2 and HCl, we use the actual bond energy (Table 4.1). Enough heat must be absorbed to break the C—H bond, whose average bond energy is 413 kJ/mol, and the Cl—Cl bond, whose actual bond energy is 243 kJ/mol. The total energy of bonds broken is $413 + 243 = 656$ kJ/mol. The heat evolved will be the sum of the average bond energy of a C—Cl bond, which is 328 kJ/mol, and the H—Cl bond energy, which is 432 kJ/mol. The total energy of bonds made is $328 + 432 = 760$ kJ/mol. Since the quantity of heat evolved is greater than the quantity absorbed, the heat of reaction is negative.

Table 4.1

Bond	Energy(kJ/mol)	Bond	Energy(kJ/mol)
H—H	436	F—F	158
H—F	563	Cl—Cl	243
H—Cl	432	Br—Br	193
H—Br	366	I—I	151
H—I	299	N≡N	946
C≡O	111	O=O	498

Bond Energies of Some Diatomic Molecules

The ΔH of the reaction must be the sum of the heat absorbed and the heat evolved.

$$\Delta H = 656 \text{ kJ/mol} - 760 \text{ kJ/mol} = -104 \text{ kJ/mol}$$

This value is quite close to the measured value of -105 kJ/mol. We can generalize this analysis as

$$\Delta H = \Sigma \text{ energies of bonds broken} - \Sigma \text{ energies of bonds made} \qquad [4.1]$$

Note that bond energies are expressed as positive numbers.

Because H, enthalpy, is a thermodynamic state function, the value of ΔH depends only on the nature of the reactants and products of the reaction. Unlike the value of the actual heat transferred, the value of ΔH does not depend on the path followed from the reactants to the products. The reason is that the definition of ΔH specifies that the reaction follows a constant-pressure path. So ΔH is the heat that would be transferred if the reaction was carried out along a constant-pressure path. This reaction, the chlorination of methane, appears quite simple but actually proceeds through a number of steps, as we shall see.

Using this approach, we can calculate the ΔH of a large number of reactions, given appropriate bond energy data.

Example 4.1

Use the data in Tables 1.3 and 4.1 to calculate ΔH of the following reactions:
(a) C_2H_5OH (ethanol) $+ HBr \rightarrow C_2H_5Br + H_2O$ (b) $C_2H_4 + H_2 \rightarrow C_2H_6$
(c) CH_3OCH_3 (dimethyl ether) $+ HBr \rightarrow CH_3Br + CH_3OH$

Solution

(a) *Step 1:* Identify the bonds broken and the bonds made. To help, it may sometimes be necessary to draw Lewis structures.

Notice that it is not necessary to include the lone pairs since we are just looking at bond breaking and bond making. Now compare the reactants and products. Ethanol, the reactant, has a C—O bond not found in the product. The product it forms has a C—Br bond not found in the reactant. Thus a C—O bond has been broken and a C—Br bond has been made. There is an H—Br bond in the reactants and none in the products. That bond is broken. There is one O—H bond in the reactants and two in the H_2O, the product. One O—H bond has been made.

Step 2: Using Tables 1.3 and 4.1, find the bond energies of the bonds that are broken or made.

Broken: C—O 358 kJ/mol   H—Br 366 kJ/mol

Made: C—Br 285 kJ/mol   O—H 463 kJ/mol

Step 3: Substitute these values into Equation [4.1]

$$\Delta H = [358 + 366] - [285 + 463] = -24 \text{ kJ/mol}$$

(b) *Step 1:*

Ethylene, the reactant, has a carbon–carbon double bond. Ethane, the product, does not. Thus a C=C bond has been broken. Ethane, the product, has a carbon–carbon single bond, ethylene does not. Thus a C—C bond has been made. (Another way to look at this change is to say a π bond has been broken. The average bond energy of a π bond is the double bond energy minus the single bond energy, which is 615 − 346 = 269 kJ/mol.) Clearly the H—H bond in H_2 has been broken. Ethylene has 4 C—H bonds; ethane has 6 C—H bonds. Two C—H bonds have been made.

Step 2:

Broken: C=C 615 kJ/mol   H—H 436 kJ/mol

Made: C—C 346 kJ/mol   2 × C—H = 2 × 413 = 826 kJ/mol

Step 3:

$$\Delta H = [615 + 436] - [346 + 826] = -121 \text{ kJ/mol}$$

(c) *Step 1*

The bonds that are broken in the reactants are an H—Br bond and a C—O bond. The bonds that are made in the products are an O—H bond and a C—Br bond.

Step 2:

Broken: H—Br 366 kJ/mol   C—O 358 kJ/mol

Made: O—H 463 kJ/mol   C—Br 285 kJ/mol

Step 3:

$$\Delta H = [366 + 358] - [463 + 285] = -24 \text{ kJ/mol}$$

4.3 | Thermochemical Relationships

As we have mentioned, bond energies do not always tell the whole story about ΔH. There can be other factors related to the structure of a chemical species that contribute to its stability or instability. How a given factor contributes to the value of ΔH will depend on whether the factor is stabilizing or destabilizing. It will depend on whether it is the reactants or the products that are stabilized or destabilized. It will depend on whether the reaction is exothermic or endothermic. A useful way to correlate these factors is with a thermochemical diagram of the type shown in Figure 4.1.

The vertical axis represents increasing enthalpy (often referred to imprecisely as energy). The horizontal axis has no meaning. Each horizontal line represents the enthalpy of a set of reactants or a set of products in a chemical reaction. If the reaction

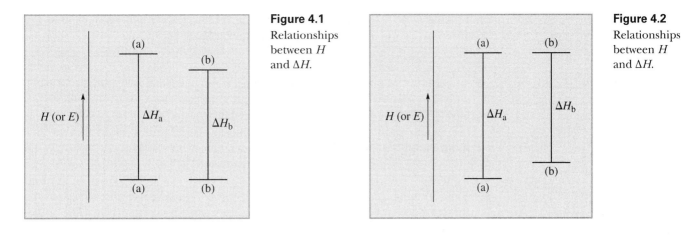

Figure 4.1
Relationships between H and ΔH.

Figure 4.2
Relationships between H and ΔH.

is exothermic, the products are lower on the vertical axis than the reactants. If the reaction is endothermic, the products are higher on the vertical axis than the reactants.

Using Bond Energies

The enthalpies represented in these diagrams may be those of actual chemical species or those of simplified imaginary chemical species that we calculate based on bond energies alone. Figure 4.1 can represent a situation where one ΔH shown is calculated from bond energies and is called ΔH_{calc}. It is not the same as the other ΔH, which is the measured ΔH, called ΔH_{obs}. If the ΔH we calculate based on bond energies is not the same as ΔH_{obs}, the difference must be due to some factor or factors in addition to bond strength that contribute to the relative stability of the reactants or products. So the difference between ΔH_{calc} and ΔH_{obs} gives us the value of those factors.

The diagram can represent any one of four possibilities.

If the reaction represented by the diagram is exothermic, the reactants are of higher energy than the products. The different vertical placement of the reactants indicates that the reason for the difference between ΔH_{calc} and ΔH_{obs} is that there is a factor in addition to bond energies influencing the stability of the reactants. If that factor is increasing the stability of the reactants, then the actual reactants are represented by the lower line (b) and $\Delta H_{obs} = \Delta H_b$. If that factor is decreasing the stability of the reactants, then they are represented by the higher line (a) and $\Delta H_{obs} = \Delta H_a$.

The difference between ΔH_{calc} and ΔH_{obs} allows us to assign a numerical value to the additional factors affecting stability. We use available data on bond energies to find ΔH_{calc}. We must be given the value of ΔH_{obs} in order to perform this calculation.

Similar logic and reasoning apply if the reaction represented by the diagram is endothermic, but everything is reversed. Now it is the products that have the factor in addition to bond energies. If that factor is a stabilizing one, then $\Delta H_{obs} = \Delta H_b$. If it is a destabilizing one, then $\Delta H_{obs} = \Delta H_a$.

Figure 4.1 shows four possible relationships. These relationships seem complex when they are described in words, but they should be easier to see taken one at a time with the aid of this kind of diagram.

Figure 4.2 shows the other four possibilities. Notice that in Figure 4.1, the two lower energy sets of compounds (reactants in endothermic reaction or products in exothermic reactions) have the same energy. In Figure 4.2 it is the two higher energy sets of compounds that have the same energy. As a result, which enthalpy difference is ΔH_{obs} and which is ΔH_{calc} switch.

If the reaction represented by the diagram is exothermic, the reactants are of higher energy than the products. The different vertical placement of the products indicates that the reason for the difference between ΔH_{calc} and ΔH_{obs} is that there is a factor in addition to bond energies influencing the stability of the products. If that factor is increasing the stability of the products, then the actual products are represented by the lower line (a) and $\Delta H_{obs} = \Delta H_a$. If that factor is decreasing the stability of the products, then they are represented by the higher line (b) and $\Delta H_{obs} = \Delta H_b$.

If the reaction represented by the diagram is endothermic, everything is reversed. It is the reactants that have the factor in addition to bond energies. If that factor is a stabilizing one, then $\Delta H_{obs} = \Delta H_a$. If it is a destabilizing one, then $\Delta H_{obs} = \Delta H_b$.

To some extent this picture is oversimplified. In all the relationships we have considered, it is either the reactants or the products whose stability differs from the one calculated using bond energies alone. But there will be reactions in which the stabilities of both the reactants and the products differ from those based on bond energies alone for more or less the same reasons. Such situations will generally be too complicated for us to analyze but will usually not cause major differences between ΔH_{calc} and ΔH_{obs}.

Resonance Energy

We have already discussed some factors that can influence stability in addition to bond strength. One of the most important is resonance (Section 1.3). A resonance hybrid is more stable than we expect on the basis of any one of its contributing structures. The extra stability is called its resonance energy. If a reactant in a chemical reaction is a resonance hybrid and the product is not, then the value of ΔH_{obs} compared to ΔH_{calc} will reflect that extra stability.

As shown in Figure 4.1 for an exothermic reaction, reactant (b) is more stable than reactant (a). Reactant (b) is the resonance hybrid, while reactant (a) is the best of the imaginary contributing structures. When we calculate ΔH_a (ΔH_{calc}), we use the energies of the bonds in this contributing structure. When we measure ΔH_b (ΔH_{obs}), the resonance energy is also included. Figure 4.1 reveals that an exothermic reaction is less exothermic for a resonance hybrid than what is predicted based on bond energies alone. The actual difference allows us to calculate the value of the resonance energy, which is always expressed as a positive number.

$$\text{Resonance energy} = |\,\Delta H_{calc} - \Delta H_{obs}\,|$$

Figure 4.1 also shows an endothermic reaction in which the product, but not the reactant, is a resonance hybrid. Such a process will be less endothermic (ΔH_b) than what is calculated (ΔH_a) from bond energies alone.

The other two combinations are shown in Figure 4.2. For an endothermic reaction, reactant (a) is more stable than reactant (b). Reactant (a) is the resonance hybrid, while reactant (b) is the best of the imaginary contributing structures. When we calculate ΔH_b (ΔH_{calc}), we use the energies of the bonds in this contributing structure. When we measure ΔH_a (ΔH_{obs}), the resonance energy is also included. Figure 4.2 reveals that an endothermic reaction is more endothermic for a resonance hybrid than what is predicted based on bond energies alone.

Figure 4.2 also shows an exothermic reaction in which the product, but not the reactant, is a resonance hybrid. Such a process will be more exothermic (ΔH_a) than what is calculated (ΔH_b) from bond energies alone.

Example 4.2

The following reaction is one in which the reactant (called 1,3-butadiene) is resonance stabilized, but the product is not.

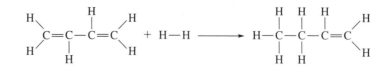

The ΔH_{obs} for this reaction is -110 kJ/mol. Find the resonance energy of 1,3-butadiene.

Solution

We must assume that the only factors that determine the value of ΔH_{obs} are the bond energies and the resonance energy. We use bond energies to find ΔH_{calc} (see Example 4.1).

Broken: C=C 615 kJ/mol   H—H 436 kJ/mol

 Made: C—C 346 kJ/mol $2 \times$ C—H $= 2 \times 413 = 826$ kJ/mol

$$\Delta H_{calc} = [615 + 436] - [346 + 826] = -121 \text{ kJ/mol}$$

Then we find the resonance energy from the difference between the two ΔH values.

$$| \Delta H_{calc} - \Delta H_{obs} | = | -121 \text{ kJ/mol} - (-110 \text{ kJ/mol}) | = 11 \text{ kJ/mol}$$

This calculation illustrates the general approach, which is to use bond energies to calculate a ΔH, which is then compared to a measured ΔH. The difference gives us some idea of the magnitude of whatever other factors are influencing the stability of the reactants or products.

This general approach can be useful, but it will usually not be very accurate except for relatively simple systems. One problem is that with the exception of diatomic molecules, the bond energies are the average of those for the given bond in many different chemical species. They are not specific for the compounds in the reaction under consideration. Another problem is that we are usually not able to isolate just one factor that influences the stability in either the reactants or the products. Most often, there is more than one and it may affect the products differently than the reactants.

Relative Stability

We can, however, compare appropriate thermochemical measurements and use a somewhat different approach to get more accurate values of quantities such as resonance energy and more important to find the relative stability of different compounds.

Figure 4.3 shows an exothermic reaction of two different reactants that form the same product. It is a specific example of one of the cases represented in a more general way in Figure 4.1.

In the diagram, reactant (b) is lower and therefore more stable than reactant (a). Because the reaction is exothermic, the product is lower on the vertical axis than the reactants. Since reactant (b) is lower than reactant (a), reactant

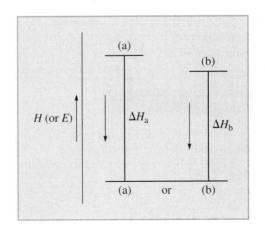

Figure 4.3
Relationships between H and ΔH for exothermic reactions.

(b) is closer to the product. So as the diagram shows, ΔH_b has a smaller numerical value than ΔH_a. The exothermic reaction of a more stable compound evolves less heat than that of a less stable compound. If the values of ΔH_b and ΔH_a have been measured, then the difference between them is the difference in stability between reactant (b) and reactant (a). Remember (and it is easily seen in the diagram, if you forget) that the more stable reactant evolves less heat in an exothermic process. This rule indicates which of the two reactants is more stable. The value of this difference is as accurate as the accuracy of the measurements. There are no issues of the kind we encountered when using average bond energies.

Figure 4.4

Relationships between H and ΔH for endothermic reaction.

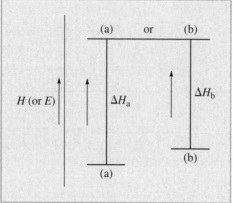

Figure 4.4 shows an endothermic reaction of two different reactants that form the same product. It is a specific example of one of the cases represented in a more general way in Figure 4.2.

In the diagram, reactant (a) is lower on the vertical axis and therefore more stable than reactant (b). Because the reaction is endothermic, the product is higher on the vertical axis than the reactants. Since reactant (a) is lower than reactant (b), reactant (a) is further away from the product. So as the diagram shows, ΔH_a has a larger numerical value than ΔH_b. The difference between them is the difference in the stability of the two reactants. The endothermic reaction of a more stable compound absorbs more heat than that of a less stable compound. This rule indicates which of the two reactants is more stable.

One of the most common applications of this approach uses combustion reactions, which are exothermic. Heats of combustion are relatively easy to measure. The products of a combustion reaction (a rapid reaction with O_2) are CO_2 and H_2O (usually liquid H_2O). The amounts of CO_2 and H_2O produced and the amount of O_2 required depend on the composition of the reactant. For example, combustion of one mole of CH_4 will produce one mole of CO_2 and two moles of H_2O, while combustion of one mole of C_2H_6 will produce two moles of CO_2 and three moles (6/2) of H_2O.

If different compounds have the same composition, they will form the same amounts of CO_2 and H_2O on combustion. Different compounds with the same composition are called isomers. Measurement of their heats of combustion is a very useful way to measure their difference in stability.

Example 4.3

There are two isomers of C_4H_{10}. Each of them will form 4 mol CO_2/1 mol C_4H_{10} and 5 mol H_2O/1 mol C_4H_{10}.

One of these isomers is called butane. It has the structure

$$
\begin{array}{cccc}
H & H & H & H \\
| & | & | & | \\
H-C-C-C-C-H \\
| & | & | & | \\
H & H & H & H
\end{array}
$$

and a heat of combustion of 2878 kJ/mol. (Remember combustion is exothermic, so the actual heat of the reaction is −2878 kJ/mol.)

The other isomer is called isobutane. It has the structure

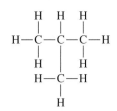

and a heat of combustion of 2869 kJ/mol. Which is the more stable isomer and by how much?

Solution

As Figure 4.3 shows, the more stable isomer has a numerically smaller heat of combustion. So isobutane is more stable than butane and the difference is reflected in the difference in their heats of combustion:

$$2878 \text{ kJ/mol} - 2869 \text{ kJ/mol} = 9 \text{ kJ/mol}.$$

An approach using average bond energies would also give this enthalpy difference because we would use the same bond energies for the bonds broken in both compounds. In each one we break 3 C—C bonds and 10 C—H bonds. Since both compounds form the same product, the ΔH_{calc} is the same for both. So the difference between them is just the difference between the ΔH_{obs} of each. Because of the large values of ΔH and the large number of bonds made and broken, a bond energy calculation is tedious. It is also unnecessary.

Sometimes using average bond energies not only is unnecessary, but gives the wrong result.

Example 4.4

Consider the combustion of the two isomers of C_2H_6O. The reaction is

$$C_2H_6O + 3O_2 \rightarrow 2CO_2 + 3H_2O$$

The two isomers are ethanol and dimethyl ether. The heat of combustion of ethanol is 1371 kJ/mol, and the heat of combustion of dimethyl ether is 1431 kJ/mol. Use these measured values to find which isomer is more stable and by how much. Repeat the calculation using average bond energies.

Solution

These measurements tell us that ethanol, which has the smaller heat of combustion, is more stable than dimethyl ether. The difference in stability is the difference in the measured heats of combustion, 1431 kJ/mol − 1371 kJ/mol = 60 kJ/mol.

The combustion of either isomer makes the same bonds, 4 C=O bonds in $2CO_2$ and 6 O—H bonds in $3H_2O$. Also 3 O=O bonds are broken in each reaction. But the other bonds broken differ between the isomers. For ethanol we break 1 C—C bond, 1 C—O bond, 1 O—H bond, and 5 C—H bonds. The total energy (in kJ/mol) required to break these bonds is

$1 \times$ C—C = 346, $1 \times$ C—O = 358, $1 \times$ O—H = 463, $5 \times$ C—H = 2065 kJ/mol

The total bond energy of the bonds that are broken in the combustion of ethanol is 3232 kJ/mol.

For dimethyl ether we break 2 C—O bonds and 6 C—H bonds. The total energy (in kJ/mol) required to break these bonds is

$$2 \times 358 + 6 \times 413 = 3194 \text{ kJ/mol}$$

These numbers indicate that breaking all the bonds in ethanol requires 38 kJ/mol more energy than breaking all the bonds in dimethyl ether. This calculation indicates that ethanol is 38 kJ/mol more stable than dimethyl ether. But the measurements indicate that ethanol is 60 kJ/mol more stable than dimethyl ether and should therefore require 60 kJ/mol more to break all its bonds. The difference calculated using bond energies does not correspond to the measured difference because of the approximations in using average bond energies. The measured difference is a much more reliable number.

When comparing the stability of different compounds or measuring a thermochemical property, it is almost always preferable to use measurements of the ΔH of reactions if they are available.

In Example 4.2 we found the resonance energy of 1,3-butadiene

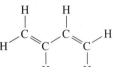

by looking at the difference between the measured heat (ΔH_{obs}) of its reaction with H_2 and the calculated heat of this reaction using average bond energies. But it is easier and more accurate to compare the measured heat of

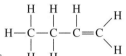

reaction of 1,3-butadiene with H_2 to that of (which is called 1-butene) with H_2. The heat of reaction of 1-butene with H_2 is -127 kJ/mol. We can take this to be the value of ΔH for the addition of H_2 to an unconjugated double bond. The ΔH for the same reaction in 1,3-butadiene, the one shown in Example 4.2, is -110 kJ/mol. This reaction is less exothermic because of the stabilization of 1,3-butadiene due to resonance (Figure 4.3). The difference between the two values is $127 - 110 = 17$ kJ/mol, which is the resonance energy of 1,3-butadiene. This value is more accurate (and easier to calculate) than the value of 11 kJ/mol we found in Example 4.2.

Once we have the value of the resonance energy, we can use it to calculate the ΔH of other reactions in which the resonance-stabilized compound reacts or is formed. When doing such calculations, you should always keep the sign convention in mind. As we have seen, processes that are energetically favorable, such as making bonds, have a negative sign when calculating ΔH and processes that are unfavorable, such as breaking bonds, have a positive sign. Thus the energy of bonds formed is subtracted and the energy of bonds broken is added. This calculation is consistent with the sign convention for ΔH. When it is negative (exothermic), the reaction is favorable; and when it is positive (endothermic), the reaction is unfavorable for all enthalpy-controlled reactions.

Similarly if we transform a compound that has resonance energy in a way that eliminates the resonance, we add the value of resonance energy in the calculation. Losing resonance energy is unfavorable. If we form a compound that is resonance stabilized, we subtract its resonance energy in the calculation. Creating resonance energy is favorable. The same reasoning can be applied to any other factor. Subtract it if it is favorable to create it or to lose it, and add it if it is unfavorable to create it or lose it.

Example 4.5

Calculate ΔH of the reaction of 1,3-butadiene with Br_2 to form the dibromide.

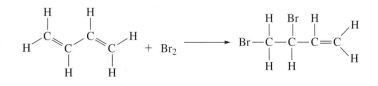

Solution

Broken: C=C 615 kJ/mol   Br—Br 193 kJ/mol

Lost: 17 kJ/mol resonance energy

Made: C—C 346 kJ/mol $2 \times$ C—Br $= 2 \times 285 = 570$ kJ/mol

$\Delta H = [615 + 193] + 19 - [346 + 570] = -91$ kJ/mol

Accurate measurements of heats of reactions can reveal subtle relationships between structure and stability. For example, we have seen that H_2 readily adds to a C=C double bond, converting it to a C—C single bond and forming two C—H bonds. The heats of these kinds of reactions, which are called hydrogenation reactions, are relatively easy to measure. Isomers of C_5H_{10} add H_2 to form C_5H_{12}. If we measure the heat of reaction of isomers of C_5H_{10} that give the same isomer of C_5H_{12}, we can compare the stability of the two C_5H_{10} isomers.

For example,

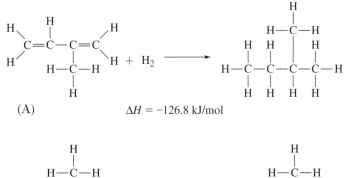

(A) $\Delta H = -126.8$ kJ/mol

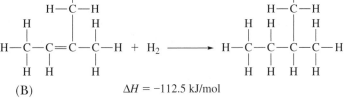

(B) $\Delta H = -112.5$ kJ/mol

The isomer in reaction B is more stable than the one in reaction A, since its heat of reaction is numerically smaller (Figure 4.3) for this exothermic reaction. The difference in stability is $126.8 - 112.5 = 14.3$ kJ/mol. This difference in stability cannot be explained with average bond energies or by resonance. One possible explanation is that a double bond with fewer carbons and more hydrogens attached (reaction A) is not as strong as a double bond with more carbons and fewer hydrogens attached (reaction B).

This example illustrates a very useful application of thermochemical measurements. By careful selection of comparison compounds, we can study the relationship between stability and a particular structural feature.

Here is another example. We found that isobutane is somewhat more stable than its isomer butane (Example 4.3). In comparing the pattern of C—C bonds in the two compounds, we notice that isobutane has one carbon bonded to three other carbons. The difference in stability could be due to that structural feature. We can test the idea that the more C—C bonds a given carbon forms, the stronger they are.

There are three isomers with the formula C_5H_{12}. They are

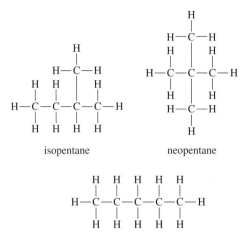

isopentane   neopentane

Based on the results for butane and isobutane, we predict that the most stable of these isomers is neopentane, followed by isopentane, and pentane as the least stable. The measured heats of combustion are neopentane 3514 kJ/mol, isopentane 3528 kJ/mol, and pentane 3536 kJ/mol. These values support our prediction. The most stable isomer is the one with the smallest numerical heat of combustion.

4.4 | Thermodynamics

Thermodynamics can be defined as the study of the relationship between energy and chemical change. We use it to provide answers to the following three questions:

1. Can a chemical or physical change be expected to occur spontaneously when one or more substances are in a given set of conditions?
2. How far can such a change proceed?
3. If a chemical or physical change occurs, what are the accompanying energy changes?

So far we have tried to answer some of these questions by considering only the enthalpy change, ΔH, as a measure of the relative stability of reactants and products. We have said that the chemical change will occur when ΔH is negative because the energy of the system decreases. It will not occur when ΔH is positive because the energy of the system increases. Because of the nature of organic reactions, most of the time, although not always, the answers we get by just considering enthalpy are sufficient for our purposes. But if we want more complete answers to these questions we must consider things from a broader thermodynamic perspective.

One basic limitation of thermodynamics must be kept in mind. Thermodynamics does not provide any information about time. It tells us that a certain chemical change is expected to occur, but it does not tell us how long we must wait for it to happen. The meaning of the word "spontaneously" in thermodynamics is somewhat different from its ordinary connotation. It has nothing to do with time. It just means that the system changes without the aid of any additional external agents.

For example, thermodynamics tells us that a mixture of CH_4 and O_2 will completely form CO_2 and H_2O, with the evolution of a great deal of heat. We do not have to do anything but wait; the process is spontaneous. But thermodynamics does not tell us that if we do not have a spark or flame, the wait will be very long, more than a human lifetime. From a practical point of view, how long it takes for a chemical change to occur is equally important. That question is addressed by kinetics, which will be discussed in Chapter 5.

Entropy

We know from our experience that not all changes can be explained with ΔH. For example, we know what happens when we put two blocks of metal, one hot and one cold, together. They eventually reach the same temperature. But, since there is no transfer of heat to or from the blocks to their surroundings, $\Delta H = 0$ for this process. Now consider the reverse process, for which $\Delta H = 0$, as well. It does not take place. If we place two blocks of metal together, each at the same temperature, one does not get warmer and the other does not get colder.

Thermodynamics defines another property called **entropy,** represented by the letter S. Entropy helps us to understand such observations. There is a change in entropy that accompanies a chemical or physical change. It is represented by the symbol ΔS. An increase in the entropy of the system, that is a positive ΔS, is favorable. A negative ΔS is unfavorable.

We have discussed some of the things that ΔH (and by extension H) measures. Most important are the relative energies of the bonds in the reactants and products. Enthalpy also measures other factors, such as resonance energy, that determine the relative energies of the components of the reacting system. A precise definition of what ΔS (and by extension S) measures can be a bit complicated. It is related to the way in which the energy of the system is distributed. But a more or less qualitative definition of entropy is very useful in organic chemistry and relatively easy to apply. Entropy is a measure of the disorganization of the system. The more disorganized the system, the greater its entropy.

A positive value of ΔS means the change causes the system to become more disorganized. More disorganization is favorable. A negative value of ΔS means that the change causes the system to become more organized. Such a change is unfavorable. Notice the sign convention for ΔS is the opposite of the one for ΔH. We can usually use common sense and experience to determine whether a given change results in more or less organization. Very often, the most important factor in bringing about a change in organization is a change in the number of particles (molecules, atoms, or ions) in the system.

Bond breaking increases disorganization. Two particles are more disorganized than one. So a change such as $Cl_2 \rightarrow 2Cl$ has a positive ΔS. But it is quite endothermic and has an unfavorable ΔH. The change is not spontaneous at room temperature, but it can be brought about by supplying heat to the system to raise its temperature. It will not occur, however, if the Cl_2 is left on its own at temperatures we normally encounter on the surface of our planet.

The reverse reaction $2Cl \rightarrow Cl_2$ results in increased organization and therefore has an unfavorable ΔS. But the reaction is quite exothermic and has a very favorable ΔH. At the temperatures we normally encounter, this reaction is spontaneous. If we bring two Cl atoms together, they form Cl_2 without the addition of heat or anything else.

Another important factor that can cause a change in organization is a phase change. Solids are more organized than liquids, and liquids are more organized than gases. So melting and evaporation of a substance are accompanied by an increase in entropy—ΔS is positive. Condensation and freezing are accompanied by a decrease in entropy—ΔS is negative. As we know, whether or not a given phase change is spontaneous depends on the temperature. Water freezes spontaneously at $-10°$ C and does not freeze at $+10°$ C. Ice melts spontaneously at $+10°$ C and does not melt at $-10°$ C.

In general, we can apply our experience from everyday life to assess relative organization and disorganization. Any process in which a relatively constrained chemical species changes to one with more freedom will have a positive ΔS. For example, consider the following change, which is brought about by heating the reactant.

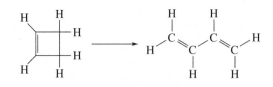

While these molecules appear to be fairly complicated, we can see that the reactant, in which the four carbon atoms are locked into a ring structure, has less freedom than the product, where there is relatively free rotation around the C—C single bond. This chemical change has a positive ΔS.

Free Energy

For processes where $\Delta H = 0$, such as the ones with the two metal blocks, we can answer the three questions we posed earlier just by considering ΔS. But most processes have a nonzero ΔH, and we must look at both ΔH and ΔS in order to decide whether the chemical or physical change we are considering will take place.

The four possible combinations of ΔS and ΔH are shown in Table 4.2.

Table 4.2

ΔH	ΔS	Process
−	+	will occur
+	−	will not occur
−	−	may occur
+	+	may occur

The first two combinations are unambiguous. In the first example, both ΔH and ΔS are favorable, so the process will occur. In the second example, both ΔH and ΔS are unfavorable, so the process will not occur. In the last two combinations, ΔH and ΔS are opposite; one is favorable and one is unfavorable. As we have said, for most organic reactions that occur under ordinary conditions, we assume that ΔH is controlling. So we assume that a process described by the third combination, in which ΔH is negative and favorable, will also occur. We assume that a process described by the fourth combination, in which ΔH is positive and unfavorable, will not occur.

There are processes, however, that are exceptions to this assumption. There are also processes that occur as predicted but still are controlled at least in part by ΔS. So when determining what the best way might be to bring about a given chemical change, we need to consider both ΔH and ΔS.

Thermodynamics defines another property called **free energy,** represented by G, which combines H and S. It is the free energy change ΔG that actually determines whether or not a process will occur. If ΔG is negative the process occurs. If ΔG is positive it does not. The sign and value of ΔG for a given chemical or physical change depend not only on the components of the system, but also on the temperature. So an unfavorable process may become favorable by changing the temperature.

We have already seen an example of such a process. At room temperature Cl_2 does not form $2Cl$. The reason is that ΔG is positive because of the positive ΔH of bond breaking. But if we raise the temperature, the reaction does take place. The reason is that as the temperature is increased, ΔS becomes more important.

One of the most important relationships in chemistry is the one that connects free energy, enthalpy, and entropy.

$$\Delta G = \Delta H - T\Delta S$$

where T is the temperature in kelvins.

The equation shows the connection between the sign of ΔG, which determines whether the process is favorable, and the signs of the ΔH and ΔS. The equation also shows why ΔS becomes more important as the temperature increases. An extreme example of the relationship between entropy, disorganization, and temperature is seen in the sun. Because the temperature is so high in the sun, there are no molecules. No chemical bond can remain intact because at these high temperatures, the $T\Delta S$ term for breaking the bond is overwhelmingly large. There are not even any atoms on the sun, just plasma. Again the $T\Delta S$ term for separating the electrons from the nuclei of atoms is overwhelmingly large.

Not only does the sign of ΔG indicate whether a chemical change will occur, but the magnitude of ΔG indicates how far that change will proceed. Another extremely important relationship in chemistry is the one that connects ΔG to the extent of reaction.

$$\Delta G = -RT\ln K$$

where K is the equilibrium constant. The more negative is ΔG, the larger is K, and the more favorable the process.

Example 4.6

Using the data in Tables 1.3 and 4.1 as necessary, predict whether or not the following gas-phase reactions at 300 K must be spontaneous, may be spontaneous or nonspontaneous, or cannot be spontaneous. Explain why.

  (a) $CH_4 + Br_2 \rightarrow CH_3Br + HBr$
  (b) $CH_3I + H_2O \rightarrow CH_3OH + HI$
  (c) $C_2H_2 + 2H_2 \rightarrow C_2H_6$
  (d) $C_2H_5Cl \rightarrow C_2H_4 + HCl$
  (e) $3CO_2 + 4H_2O \rightarrow C_3H_8 + 5O_2$

Solution

This type of analysis and prediction starts with an assessment of the sign of ΔH and of ΔS. We use average bond energies to find the sign and magnitude of ΔH, and we use general ideas about disorganization to find the sign of ΔS.

  (a) Bonds broken: C—H 413, Br—Br 193 = 606 kJ/mol
      Bonds made: C—Br 285, H—Br 366 = 651 kJ/mol
      $\Delta H = 606 - 651 = -45$ kJ/mol
      There are two reactant molecules and two product molecules. ΔS should be zero or very close to zero.
      The reaction is spontaneous because it is exothermic with no significant entropy change.

  (b) Bonds broken: C—I 218, O—H 463 = 681 kJ/mol
      Bonds made: C—O 358, H—I 299 = 657 kJ/mol
      $\Delta H = 681 - 657 = + 24$ kJ/mol
      There are two reactant molecules and two product molecules. ΔS should be zero or very close to zero.
      The reaction is not spontaneous because it is endothermic with no significant entropy change.

  (c) Bonds broken: C≡C 812, 2 × H—H 2 × 436 = 1684 kJ/mol
      Bonds made: C—C 346, 4 × C—H 4 × 413 = 1998 kJ/mol
      $\Delta H = 1684 - 1998 = -314$ kJ/mol
      There are three reactant molecules and one product molecule. ΔS will therefore be negative and unfavorable.
      But this reaction is quite exothermic. The quantity of heat evolved corresponds to that for the formation of a reasonably strong bond.
      The reaction is very likely to be spontaneous because of the large negative ΔH.

  (d) Bonds broken: C—Cl 328, C—H 413 = 741 kJ/mol
      Bonds made: C—C π bond, 615 – 346, H—Cl 432 = 701 kJ/mol

$$\Delta H = 741 - 701 = +40 \text{ kJ/mol}$$

There is one reactant molecule and two product molecules. ΔS will therefore be positive and favorable.

It is not clear whether this reaction is spontaneous or not. We have said that most organic reactions tend to be controlled by ΔH. So this reaction is unlikely to be spontaneous. But it is close enough that a measurement is needed to answer the question. If we wish to carry out such a reaction, in which the entropy is favorable, we may want to use heat.

(e) Since this reaction is the reverse of the combustion of propane, a highly exothermic reaction, it must be highly endothermic. ΔH is positive. Since seven molecules are forming six molecules, ΔS is negative. This reaction cannot be spontaneous at any temperature.

4.5 | Thermodynamics and Chemical Reactions

One of the most important goals of organic chemistry is to understand the details at the atomic/molecular level of how and why a given reaction takes place. The details of how are called a reaction mechanism. Thermodynamics can be a useful aid in formulating and understanding reaction mechanisms. Thermodynamics can be an even more powerful tool in helping us to predict and understand chemical behavior.

The reactions of alkanes with halogens provide a very good example of how we can apply thermodynamic principles to explain differences in the chemical behavior of related systems. The reactions can be represented as

$$CH_4 + X_2 \rightarrow CH_3X + HX$$

where X is any of the halogens—F, Cl, Br, or I—and CH_4 represents any alkane. Based on substantial experimental evidence, we believe that formation of the products from the alkane occurs in two steps. First is the attack of a halogen atom $X\bullet$ (initially generated from X_2), which abstracts an H atom from the alkane:

1. $CH_4 + X\bullet \rightarrow CH_3\bullet + HX$

The products of this step are HX, an ordinary molecule, and a very reactive species in which a carbon atom has only 7 electrons. Such a species, which has an odd number of electrons, is called a free radical.

In the second step, the free radical reacts with X_2 to form an alkyl halide and a halogen atom.

2. $CH_3\bullet + X_2 \rightarrow CH_3X + X\bullet$

The $X\bullet$ formed in the second step then reacts with another molecule of alkane in a repeat of the first step.

A thermodynamic analysis of these two steps is straightforward. In each step a bond is being broken and a bond is being formed. We can find ΔH from the relevant bond energies. In each step there are two chemical species as reactants and two chemical species as products. We expect ΔS to be very close to 0 and the reactions to be enthalpy controlled.

We can use the bond energy data in Tables 1.3 and 4.1 to find ΔH for each step for each halogen. The relevant data are repeated in Table 4.3.

Table 4.3	Bond Energies for Halogenation of Alkanes				
C—X Bond	Energy(kJ/mol)	X—X Bond	Energy(kJ/mol)	H—X Bond	Energy (kJ/mol)
C—F	489	F—F	158	H—F	563
C—Cl	328	Cl—Cl	243	H—Cl	432
C—Br	285	Br—Br	193	H—Br	366
C—I	218	I—I	151	H—I	299
C—H	439 in CH_4		423 in C_2H_6		

We can see from these bond energies that the second step is always exothermic and favorable. In this step an X—X bond is broken and a C—X bond is formed. The C—X bond is stronger than the X—X bond for every halogen. The differences (which are the ΔHs of the second step) range from 158 kJ/mol − 489 kJ/mol = −331 kJ/mol for fluorine to 151 kJ/mol − 218 kJ/mol = −67 kJ/mol for iodine.

It is therefore the first step that plays an important role in determining the course of the overall reaction for different halogens. In this step a C—H bond is being broken and an H—X bond is being formed. In the case of fluorine the strong H—F bond causes this step to be quite exothermic. ΔH = 439 kJ/mol − 563 kJ/mol = −124 kJ/mol for CH_4. It is even more exothermic for other alkanes, all of which have weaker C—H bonds than CH_4. In the case of chlorine, the reaction will be very slightly exothermic or endothermic depending on the alkane. For bromine and iodine, the first step is clearly endothermic.

We can make some predictions from this analysis. Because both steps are quite exothermic and favorable for fluorine, we predict that the reaction of alkanes with fluorine should proceed quite readily. In fact when F_2 is mixed with an alkane, there is a danger of explosion because the reaction is so vigorous and exothermic. Unless carefully controlled, reactions of alkanes with F_2 are not very useful.

In the case of chlorine, the first step is at best only slightly exothermic, depending on the nature of the alkane and the C—H bond that is broken when the Cl• abstracts an H atom. But the reaction should proceed fairly easily because the second step is quite exothermic, ΔH = 243 kJ/mol − 328 kJ/mol = −85 kJ/mol. Chlorination of alkanes does proceed quite readily. But, unlike fluorination, it can easily be carried out under conditions where it can be controlled.

The second step in bromination is slightly more exothermic than for chlorination. ΔH = 193 kJ/mol − 285 kJ/mol = −92 kJ/mol. The first step, however, is endothermic; the two steps taken together are exothermic. The second step should be sufficiently exothermic to make bromination of alkanes a feasible reaction, which it is.

The reaction of alkanes with I_2 appears different. The first step is quite endothermic, but the second step is only somewhat exothermic. We can predict correctly that the second step is not sufficiently favorable to overcome the quite unfavorable first step. Iodination of alkanes is not a feasible reaction.

We can see that the trend in the values of ΔH for the first step directly follows the trend in the strength of the H—X bond, which is the bond made in the first step. As the H—X bond weakens from H—F to H—I, the corresponding halogen becomes less and less reactive.

These thermodynamic parameters provide us insight into another aspect of the reaction of alkanes with halogen. The reaction of methane, CH_4, or ethane, C_2H_6,

$$
\begin{array}{cc}
\text{H} & \text{H H}\\
| & | \ |\\
\text{H—C—Cl} & \text{H—C—C—Cl}\\
| & | \ |\\
\text{H} & \text{H H}
\end{array}
$$

with a halogen produces one product, for example: H   or   H H   .

But for any larger alkane, more than one product is almost always possible. For example, consider the halogenation of propane.

The relative amounts of the products that form (called the product distribution) are determined in the first step of the reaction. They depend on which H

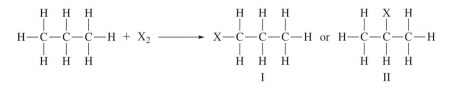

atom is abstracted from the alkane by the halogen atom. Propane has a total of eight hydrogens, six of one kind (attached to the end carbons) and two of another kind attached to the middle carbon. If one of the six end hydrogens is abstracted by the halogen atom, then product I is obtained. If one of the middle two hydrogens is abstracted, then product II is obtained. It would appear that the ratio of product I to product II should be 6:2. This probability is based on the number of each of the two kinds of hydrogens.

But there is another very important factor to consider. The strengths of the C—H bonds with different types of hydrogens are not the same. As a general rule, the fewer H atoms that are bonded to a given carbon, the weaker are the C—H bonds. The abstraction of the hydrogen with the weaker bond will be thermodynamically favored. Breaking a weaker bond makes the reaction more exothermic or less endothermic.

Thus the product distribution also depends on how selective the halogen atom is when it abstracts a hydrogen atom. If the halogen atom is selective, there will be more of the product resulting from abstraction of the hydrogen of the weaker C—H bond. For a given C—H bond, how much more will depend on just how selective is the halogen. Under a given set of conditions, the selectivity of the halogen is directly related to the ΔH of the first step. The more exothermic or less endothermic the step, the less selective is the halogen.

We can correctly predict that because step 1 for fluorine is quite exothermic, fluorine shows virtually no selectivity around room temperature, and the product distribution simply reflects the numbers of each kind of hydrogen. Step 1 for chlorine can be slightly exothermic or slightly endothermic depending on the strength of the C—H bond of the H it is abstracting. We observe only very slight selectivity around room temperature for chlorine. Step 1 for bromine is clearly endothermic, and as a result bromine is quite selective around room temperature.

These halogenation reactions are one of many examples of chemical reactions for which a consideration of thermodynamic factors helps us to understand how and why the reactions occur the way they do. When explaining or predicting chemical behavior, you should always be sure that your explanations are consistent with thermodynamics.

Exercises

4.1 Which of the following processes are exothermic?

   (a) Freezing of water at $0°\,C$; **(b)** freezing of water at $-10°\,C$; **(c)** conversion of glucose to carbon dioxide and water; **(d)** evaporation of water at $110°\,C$; **(e)** electrolysis of water to form H_2 and O_2; **(f)** walking up a flight of steps.

4.2 Predict which of the following reactions are likely to have a positive ΔH without actually doing a calculation using bond energies. Justify your predictions.

   (a) $N_2O_4 \rightarrow 2NO_2$

   (b) $C_2H_6 \rightarrow C_2H_4 + H_2$

   (c) $HCl + NaOH \rightarrow H_2O + NaCl$

   (d) $CO_2 + 2H_2O \rightarrow CH_4 + 2O_2$

   (e) $C_2H_4 + Cl_2 \rightarrow C_2H_4Cl_2$

   (f) $2H_2O \rightarrow H_2O_2 + H_2$

4.3 For each of the following pairs of compounds, predict which one has the greatest heat of combustion in kJ/mol and explain why.

   (a) C_3H_8 or C_2H_6

   (b) $CH_3CH\!=\!CH\!-\!CH\!=\!O$ or $CH_2\!=\!CHCH_2CH\!=\!O$

4.4 The reaction of N_2 and F_2 in the gas phase forms NF_3. The ΔH of this reaction is -114 kJ/mol NF_3. Use the data in Table 4.1 to find the average bond energy of an N—F bond in NF_3.

4.5* The reaction of Br_2 and F_2 can form BrF_3 or BrF_5 in the gas phase. The ΔH of the reaction for the formation of 1 mol BrF_3 is -256 kJ/mol, and for the formation of 1 mol of BrF_5 it is -429 kJ/mol. Use the data in Table 4.1 to find the average bond energy of the Br—F bond in these two molecules. Explain why the values are different.

4.6 All four halogens can add to C_2H_4 to form $C_2H_4X_2$. Calculate ΔH for each of these four reactions. Identify the most important factor in determining the favorability of reaction.

4.7 We can imagine the conversion of CH_3CH_2Br to CH_3CH_2Cl by a one-step reaction or a two-step pathway:

   1. $CH_3CH_2Br + HCl \rightarrow CH_3CH_2Cl + HBr$

   or

   2. $CH_3CH_2Br \rightarrow CH_2{=\!\!=}CH_2 + HBr$

     $CH_2{=\!\!=}CH_2 + HCl \rightarrow CH_3CH_2Cl$

Use the data in Tables 1.3 and 4.1 to calculate ΔH for the one-step process and the two-step process. Compare the values and explain the result.

4.8 Bond energies are sometimes measured by measuring the heat of atomization, which is the heat required to convert a molecule into the separate atoms of which it is composed. The heat of atomization of CH_4 is 1756 kJ/mol, and the heat of atomization of CH_2F_2 is 1876 kJ/mol. Use these data to calculate the average bond energy of the C—F bond in CH_2F_2.

4.9 The ΔH of the reaction between C_2H_2 and HCl to form vinyl chloride, $CH_2{=\!\!=}CHCl$, is -149 kJ/mol. Use the data in Tables 1.3 and 4.1 to estimate the resonance energy of vinyl chloride.

4.10* Find the ΔH of the reaction $N_2(g) + 1/2 O_2(g) \rightarrow N_2O(g)$ using the data in Tables 1.3 and 4.1 and the following additional data. The resonance energy of N_2O is 88 kJ/mol. The N=N bond energy is 418 kJ/mol and the N=O bond energy is 607 kJ/mol.

4.11 The ΔH of the reaction $3C(g) + 2H_2(g) + 1/2 O_2(g)$ to form acrolein, $CH_2{=\!\!=}CH{-}CH{=\!\!=}O(g)$, is -2323 kJ/mol. Use the data in Tables 1.3 and 4.1 to find the resonance energy of acrolein. Compare this value to that calculated for 1,3-butadiene in Example 4.2 and account for the difference.

4.12 Both propene   and cyclopropane   add H_2 to form propane, C_3H_8. The heat of hydrogenation of propene is 127 kJ/mol, and the heat of hydrogenation of cyclopropane is 242 kJ/mol. Suggest a reason for this difference.

4.13* Use the result of Exercise 4.12 and the data in Tables 1.3 and 4.1 to calculate the ΔH of the reaction Cyclopropane $+ Cl_2 \rightarrow ClCH_2CH_2CH_2Cl$ in the gas phase.

4.14* Use the result of Exercise 4.12 and the data in Tables 1.3 and 4.1 to calculate the ΔH of the reaction $C_2H_4 + CH_2N_2 \rightarrow$ cyclopropane $+ N_2$. (Assume that the NN bond in CH_2N_2 is half way between an NN double bond and an NN triple bond; and that the CN bond is halfway between a CN single bond and a CN double bond. See Exercise 1.16.) The bond energy of the N=N bond is 418 kJ/mol.

4.15 Given the following heats of combustion (in kJ/mol) of alkanes in the gas phase: propane C_3H_8 = 2220; butane C_4H_{10} = 2878; pentane C_5H_{12} = 3535; and hexane C_6H_{14} = 4192, predict the heat of combustion of heptane, C_7H_{16}.

4.16 Predict the sign of ΔS for each of the following reactions. Justify your answers.

   (a) $3H_2(g) + N_2(g) \rightarrow 2NH_3\ (g)$

   (b) $CH_4(g) + 4Cl_2\ (g) \rightarrow CCl_4(l) + 4HCl(g)$

   (c) $CH_4(g) + Br_2\ (l) \rightarrow CH_3Br(g) + HBr(g)$

   (d) $H_2(g) + F_2(g) \rightarrow 2\ HF\ (g)$

   (e) $CH_2ICH_2I(g) \rightarrow CH_2{=\!\!=}CH_2(g) + I_2(s)$

4.17 Which of the two reactions of Exercise 4.12 has the greater ΔS?

4.18 Using the data in Tables 1.3 and 4.1 as necessary, predict whether or not the following gas-phase reactions at 300 K must be spontaneous, may be spontaneous or nonspontaneous, or cannot be spontaneous. Explain why.

   (a) $CH_3Br + NH_3 \rightarrow CH_3NH_2 + HBr$

   (b) $C_2H_2 + 5/2 O_2 \rightarrow 2CO_2 + H_2O$

*More challenging questions.

(c) $C_2H_6 + I_2 \rightarrow C_2H_5I + HI$

(d) $C_2H_4 + F_2 \rightarrow C_2H_4F_2$

(e) $C_2H_2Cl_2 \rightarrow C_2H_2 + Cl_2$

4.19 A chemistry student was asked to propose a mechanism for the gas-phase reaction

$$CH_3OH + HI \rightarrow CH_3I + H_2O$$

and suggested a number of possibilities. Do a thermodynamic assessment of each of the following suggested mechanisms to decide its relative merit.

(a) $CH_3OH \rightarrow CH_3{}^{\bullet} + HO^{\bullet}$

   $HO^{\bullet} + HI \rightarrow H_2O + I^{\bullet}$

   $I^{\bullet} + CH_3{}^{\bullet} \rightarrow CH_3I$

(b) $CH_3OH + HI \rightarrow I^{\bullet} + CH_3{}^{\bullet} + H_2O$

   $I^{\bullet} + CH_3{}^{\bullet} \rightarrow CH_3I$

(c) $CH_3OH + HI \rightarrow CH_3I + H_2O$

Chapter 5
Chemical Kinetics

As we have seen in Chapter 4, thermodynamics allows us to predict whether a chemical change should happen. But from a practical point of view that is not enough information. We also need to know whether the change actually will happen. Thermodynamics tells us whether a given chemical change is possible but provides no information at all about time. Chemical kinetics tells us how much time is needed for that chemical change to occur. We cannot get complete and necessary information about a chemical change without applying both thermodynamics and kinetics.

Chemical kinetics seeks to answer three questions:

1. At what rate does a chemical system undergo change under a given set of conditions?
2. What effect will a change in conditions have on the rate at which a chemical change occurs?
3. What information do the answers to the first two questions provide about the details of how the chemical change takes place?

While the answers to the first two questions are needed to answer the third question, it is the third one that is most important in organic chemistry. The details of how a chemical change takes place are called the reaction mechanism. The study of reaction mechanisms is one of the major emphases of organic chemistry. Both thermodynamics and kinetics must be brought to bear in order to explain and understand reaction mechanisms.

Kinetics measures the rate of a chemical reaction. We are familiar in everyday life with expressions that define a rate. We drive at 100 km/hour, spend 10 cents/minute for Internet access in a café, read 20 pages/hour of organic chemistry, and drink 8 ounces of water/10 minutes. Each of these expressions describes a change that takes place in a given time interval. The rate of a chemical reaction, similarly, describes a change in a given time interval. The change is usually in the amount of one of the reactants or products of the chemical reaction.

5.1 | Reaction Rate and Concentration

The rate of a reaction is almost always related to the concentration (or pressure for gas-phase reactions) of one or more substances in the reacting system. One important goal of kinetics is to measure the exact relationship between concentration and the rate of a reaction for a given system. With this information, we are able to carry out the reaction in the most practical, efficient way. And, with this information, we can often gain valuable insights into the mechanism of the reaction. It is important to remember that *the relationship between rate and concentration for a given reaction or type of reaction can be established only by doing the measurement in the laboratory.*

The relationship between rate and concentration is called a **rate law**. Fortunately, most reactions obey one of a few relatively simple rate laws. A rate law is an algebraic equation that expresses the proportionality between the rate of a reaction and the concentration of one or more species in the system, almost always reactants. Here is a simple example. The rate of the reaction

$$2H_2O_2 \rightarrow 2H_2O + O_2$$

has been studied and found to be directly proportional to the concentration of H_2O_2. The equation is

$$Rate = k[H_2O_2],$$

where k is a proportionality constant called the specific rate constant.
Another example is the reaction

$$CH_3Br + I^- \rightarrow CH_3I + Br^-$$

whose rate is found to be proportional to the concentrations of both reactants. The rate law is

$$Rate = k[CH_3Br][I^-]$$

It is very important to remember that the rate law cannot be derived just by looking at the chemical reaction. For example, a reaction that appears simple such as

$$H_2 + Br_2 \rightarrow 2HBr$$

follows the rate law

$$Rate = k\,[H_2][Br_2]^{1/2}$$

under certain conditions.
The two central features of a rate law are the identity of the reactants whose concentrations determine the rate and the exponents of the concentration terms. The exponents define what is called the **order** of the reaction. In the rate law for the decomposition of hydrogen peroxide, the exponent on the $[H_2O_2]$ is 1. This rate law is called a first-order rate law, and the reaction is called a first-order reaction.

In the rate law for the reaction of CH_3Br and I^-, there are two concentration terms. The exponent on each is 1. This rate law is called a two-term second-order rate law, and the reaction is a second-order reaction. It is first order in $[CH_3Br]$ and first order in $[I^-]$.

There are also one-term second-order rate laws. For example, the rate law for the reaction $2NO(g) \rightarrow N_2(g) + O_2(g)$ is

$$Rate = k[NO]^2$$

The third rate law is more complicated. It is a two-term rate law, first order in $[H_2]$ and one-half order in $[Br_2]$. It can be called a three-halves order reaction.

Example 5.1

Write rate laws corresponding to the following measurements for the reaction

$$A + B + C \rightarrow products$$

and describe the rate laws in words.
(a) When [A] doubles, the rate doubles; when [B] doubles, the rate quadruples; when [C] doubles, the rate is unchanged.
(b) When [A] doubles, the rate increases by a factor of 8. When [B] or [C] doubles, there is no change in the rate.
(c) When [A] doubles, the rate doubles; when [B] doubles, the rate doubles; when [C] doubles, the rate decreases by ½.
(d) When [A], [B], and [C] each double, the rate increases by a factor of 8.

Solution

(a) Rate $= k[A][B]^2$; first order in A, second order in B, third order overall.

(b) Since $2^3 = 8$,
rate $= k[A]^3$, third order in A, third order overall.

(c) A decrease in rate with an increase in the concentration of a reactant means that the concentration term for that reactant will have a negative exponent in the rate law. Hence, rate $= k[A][B][C]^{-1}$, first order in A, first order in B, minus first order in C. When there is a negative exponent, we usually do not designate an overall order.

(d) It is not possible to write the rate law from this information, since the increase in rate could be due to the increase in the concentration of just one of the reactants (third order in that one) or of two of them (second order in one and first order in one) or all three (first order in all three).

We can do simple proportion calculations to find rates given concentrations or vice versa.

Example 5.2

(a) At a certain temperature, the rate of decomposition of H_2O_2 to H_2O and O_2 is 0.010 M/min when its concentration is 1.0 M. What will the rate of decomposition be when its concentration is 0.75 M?

(b) Carry out the same calculation for the reaction of $NO(g)$ to form $N_2(g)$ and $O_2(g)$ given that the rate is 0.010 M/min when its concentration is 1.0 M.

Solution

(a) Since the rate is directly proportional to $[H_2O_2]$

$$Rate_1/Conc_1 = Rate_2/Conc_2$$

$$(0.010 \text{ M/min})/(1.0 \text{ M}) = Rate_2/(0.75 \text{ M})$$

$$Rate_2 = 0.0075 \text{ M/min}$$

(b) Since the rate is directly proportional to $[NO]^2$

$$Rate_1/Conc_1^2 = Rate_2/Conc_2^2$$

$$(0.010 \text{ M/min})/(1.0 \text{ M})^2 = Rate_2/(0.75 \text{ M})^2$$

$$Rate_2 = 0.0056 \text{ M/min}$$

In general chemistry we carry out many calculations using kinetic data and rate laws to find such things as the actual rate of a reaction at a given time or the rate law from rate measurements. In organic chemistry we generally do not carry out such calculations. But we will still use rate laws. Perhaps more than any other kinetic parameter, the order of a reaction and the order of each of the reactants provide information about the mechanism of a reaction.

5.2 | Rate Laws and Reaction Mechanisms

A reaction mechanism is a description on a molecular level of all the changes that the reactants undergo to form the products in a chemical reaction. Ideally, a reaction mechanism will describe the movement of the electrons and nuclei of the reactants in a continuous way along the pathway to the products. But often our descriptions of reaction mechanisms are less detailed and simply present a sequence of steps.

Most chemical changes do proceed by a sequence of steps. For example, kinetic studies indicate that the conversion of ozone (O_3) to oxygen (O_2)

$$2O_3 \rightarrow 3O_2$$

does not proceed in one step. It is believed to proceed in two steps:

1. $O_3 \rightleftharpoons O_2 + O$
2. $O_3 + O \rightarrow 2\,O_2$

We can call these two steps the reaction mechanism for the conversion of O_3 to O_2. Their sum is the overall reaction.

There are several important features of a reaction mechanism:

1. The sum of all the steps in a reaction mechanism must be the overall reaction.
2. Each step in this or any mechanism is a discrete chemical reaction called an **elementary reaction**. An elementary reaction is a single-step reaction. It occurs exactly as it is written, as opposed to many chemical reactions that are complex reactions and occur in two or more steps.
3. In the first elementary reaction here, one O_3 molecule breaks into two pieces, an O_2 molecule and an O atom, exactly as written. In the second elementary reaction, one O_3 molecule reacts with one O atom to form two molecules of O_2, exactly as written.
4. The overall reaction does not occur as written. Two O_3 molecules do not react with each other.

Molecularity is a term that is applied only to elementary reactions. It gives us the rate law and the reaction order for the elementary reaction. The molecularity of an elementary reaction is the number of reactant molecules (or atoms or ions) that come together to form the products. There is one reactant molecule in the first elementary reaction of the ozone mechanism, so we say it is unimolecular. Reaction 2 has two species, a molecule and an atom. We say it is bimolecular. There are also a very few elementary reactions known in which three reactants come together. These termolecular elementary reactions are rarely encountered. Elementary reactions with a molecularity of four or greater are unknown. (Why?)

The reaction order of an elementary reaction is equal to its molecularity. Therefore we can write its rate law directly from its equation, exactly what we *cannot* do for a reaction that occurs in more than one step. A unimolecular reaction such as reaction 1 follows a first-order rate law. In this case it is

$$\text{Rate} = k_1[O_3]$$

It is common practice to add a subscript indicating the step number in the mechanism to k, the specific rate constant.

A bimolecular reaction, such as reaction 2, follows a second-order rate law. In this case it is a two-term rate law.

$$\text{Rate} = k_2[O_3][O]$$

An elementary reaction such as

$$2Cl \rightarrow Cl_2$$

follows a one-term, second-order rate law.

$$\text{rate} = k[Cl]^2$$

We know the formation of Cl_2 in this way is an elementary reaction because nothing more than the formation of a bond between two atoms is taking place. Any reaction in which the only change is one bond being formed or one bond being broken is an elementary reaction.

In any reaction mechanism of two or more steps, one of the products of an early step will not be a final product of the reaction. In the mechanism for the decomposition of O_3, the O that forms in the first step is not one of the products. It is a reactant in the second step and is consumed. Such a species is called an **intermediate**. In many such reactions, the intermediate is a relatively high-energy species with a short lifetime.

One common high-energy intermediate is a carbocation, in which a carbon atom has only 3 bonds and a total of 6 electrons. It has a positive charge as a result. Other high-energy intermediates are carbanions, which are anions in which a carbon atom has a negative charge because it has a completed octet but only three bonds. Free radicals are the third important type of high-energy intermediates. A free radical is a species with an odd number of electrons. The odd electron is generally represented by a dot. Such a species must have an atom with an incomplete octet. If the atom is carbon, it will have only three bonds. Electron bookkeeping reveals that a free radical is neutral. Figure 5.1 shows some simple examples of these high-energy intermediates.

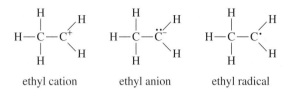

ethyl cation   ethyl anion   ethyl radical

Figure 5.1
High-energy intermediates

When we consider an overall reaction, it may be an elementary reaction or it may be a complex reaction, the result of two or more elementary reactions. Kinetics can sometimes help us to decide whether a given reaction is elementary or complex. For example, if the reaction

$$2O_3 \rightarrow 3O_2$$

were an elementary one, it would have a second-order, one-term rate law. If it does not, then it cannot be elementary. For this reaction, measurements indicate that the rate law is not second order, so the reaction must be complex. This rule is general. A reaction cannot be elementary if its rate law is not the one for the elementary reaction.

But the opposite is not true. A reaction may be complex even if its rate law is that of the elementary reaction. For such reactions, we must go beyond kinetic measurements in order to decide. Sometimes it is very difficult or even impossible to decide whether a given reaction that follows the rate law of the elementary reaction is indeed elementary or complex. In general it is not possible to prove that a given mechanism is correct. But it is possible to prove that a certain mechanism is incorrect. If the measured rate law of a reaction is inconsistent with the proposed mechanism, the mechanism is wrong.

In order to determine whether a proposed mechanism might be correct, we need to be able to derive its corresponding rate law. If the derived rate law is the same as the experimentally measured one, then the proposed mechanism *might* be correct. If it is different, then the mechanism is wrong.

It can be difficult to derive the rate law required by a proposed mechanism unless the mechanism has what is called a **rate-determining step**. When one elementary reaction in a complex mechanism proceeds at a much slower rate than any of the others, the slow elementary reaction is the rate-determining step. The rate of this step determines the overall rate of a complex reaction, no matter whether the rate-determining step comes first, last, or in the middle of the series of elementary reactions. The other steps do not directly influence the rate or the rate law of the complex reaction, but may do so indirectly, as we shall see.

The rate and rate law of the complex reaction are the rate and rate law of the rate-determining step. The rate and rate law of the complex reaction are not influenced by steps that occur after the rate-determining step. The simplest case is the one where the rate-determining step is the first step. The rate and rate law for the overall reaction just come directly from the first step. For example, the gas-phase reaction

$$NO_2 + CO \rightarrow NO + CO_2$$

is believed to occur by the following two-step mechanism:

1. $NO_2 + NO_2 \rightarrow NO + NO_3$
2. $NO_3 + CO \rightarrow NO_2 + CO_2$

in which the first step is rate determining. The observed rate law for this reaction is

$$Rate = k_1[NO_2]^2$$

This observed rate law tells us two things. First, the overall reaction is not an elementary reaction, but must proceed in two or more steps. Second, the proposed mechanism *may* be correct.

We might also consider whether the proposed mechanism makes chemical sense. One of the products of the first step is NO_3, a high-energy intermediate, not an ordinary molecule. It seems reasonable that formation of such an unstable product might be slow. This high-energy species is one of the reactants in the second step, so it is reasonable for the second step to be much faster.

When the rate-determining step is not the first step, the derivation of the rate law may be more complicated. The rate law that we derive for a proposed reaction mechanism must be one that we can check by experiment. The experimental determination of a rate law requires the measurement of how the rate changes with changes in the concentrations of the components of the chemical system. Because the determination of a rate law requires the measurements of concentrations, it can include only species whose concentrations we can measure. We can measure concentrations of only stable chemical species, such as the reactants and on rare occasions the products.

A mechanism in which the first step is not rate determining often begins with the formation of an intermediate, whose concentration we cannot measure. This intermediate will then be one of the reactants in the rate-determining elementary reaction. In order to find the rate law for such a mechanism, we must find the relationship between the concentration of this intermediate and the concentration of the substances in the system that we can measure. This relationship depends on how and from what the intermediate is formed. Very often intermediates are formed from one or more reactants by what is called a fast pre-equilibrium. They are formed rapidly and reversibly, and the concentration of the intermediate will depend on the value of the equilibrium constant for the pre-equilibrium.

For example, under certain conditions, treatment of ethyl alcohol, C_2H_5OH, with HBr, which is a strong acid, forms ethyl bromide, C_2H_5Br (Section 4.2). The overall reaction is

The mechanism for this reaction is believed to be

1. $C_2H_5OH + H^+ \rightleftharpoons C_2H_5OH_2^+$ fast
2. $C_2H_5OH_2^+ + Br^- \rightarrow C_2H_5Br + H_2O$ slow

The second step is rate determining. Its rate law is

$$Rate = k_2[C_2H_5OH_2^+][Br^-]$$

The $C_2H_5OH_2^+$ is an intermediate whose concentration cannot readily be measured. But its concentration is determined by the starting concentrations of the reactants, C_2H_5OH and H^+ (from HBr), and the equilibrium constant K for the first step, which is a Brønsted acid–base reaction that forms the conjugate acid of ethanol. It is

$$K = [C_2H_5OH_2^+]/[C_2H_5OH][H^+]$$

and

$$[C_2H_5OH_2{}^+] = K[C_2H_5OH][H^+]$$

We can substitute this expression for $[C_2H_5OH_2{}^+]$ in the rate law for step 2

$$\text{Rate} = k_2K[C_2H_5OH][H^+][Br^-]$$

to obtain a rate law in terms of measurable concentrations. The rate law for this reaction is third order, consistent with experiment.

It can be much more difficult to propose a mechanism from a measured rate law. An understanding of structure and chemical behavior can help.

For example, isopropyl alcohol, C_3H_7OH, also reacts with HBr and forms isopropyl bromide, C_3H_7Br. The overall reaction is

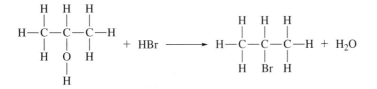

It appears very similar to the reaction of ethyl alcohol. But kinetic measurements reveal that this reaction follows a different rate law. It follows a second-order rate law.

$$\text{Rate} = k[C_3H_7OH][H^+]$$

This result may surprise you. The rate of the reaction does not depend on the $[Br^-]$. The rate-determining step must therefore take place before Br^- participates. After some study of organic chemistry, the mechanism will seem relatively straight forward. The first step is the same as it is for ethanol. It is a fast pre-equilibrium that is a Brønsted acid–base reaction that forms the conjugate acid of isopropyl alcohol.

1. $C_3H_7OH + H^+ \rightleftharpoons C_3H_7OH_2{}^+$ fast

The rate-determining step is next, and Br^- is not one of the reactants in this step. It is simply the dissociation of the conjugate acid of isopropyl alcohol into a carbocation and water.

2.

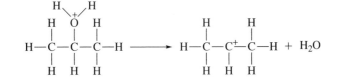

The final step, which occurs after the rate-determining step, has no effect on the rate. It is

3.

You may be wondering why the mechanism changes and why we believe that the rate-determining step is formation of a carbocation. These are the kinds of questions that you will be able to answer after some study of organic chemistry.

Knowing the rate law may not always be enough to decide between two possible mechanisms. For example, the reaction of chloroform, $CHCl_3$, with OH^- in aqueous solution forms H_2O, Cl^-, and CCl_2. A reasonable mechanism begins with an acid–base reaction:

1. $CHCl_3 + OH^- \rightleftharpoons H_2O + CCl_3{}^-$

followed by the loss of Cl^- from the intermediate anion.

2. $CCl_3^- \rightarrow Cl^- + CCl_2$

If the first step is rate-determining, the rate law is second order overall, first order in each of the reactants:

$$\text{Rate} = k_1[CHCl_3][OH^-]$$

If the second step is rate determining then

$$\text{Rate} = k_2[CCl_3^-]$$

and

$$[CCl_3^-]/[CHCl_3][OH^-] = K$$

$$[CCl_3^-] = [CHCl_3][OH^-]K$$

This mechanism also requires a rate law that is second order overall, first order in each of the reactants. So we would need another approach to decide between these two possible mechanisms.

5.3 | The Transition State

As we know, chemical reactions take time to occur and generally proceed faster as the temperature increases. Every elementary reaction in a mechanism takes time, even if it is not the slowest or rate-determining step. There are different explanations for these fundamental observations. Some of them provide important insights into how chemical changes occur. To some extent an explanation of why a particular reaction takes time depends on the particular reaction.

For example, one reason that a bimolecular elementary reaction takes time is because the two reacting species have to find each other and collide. So even a reaction such as

which is very favorable energetically still cannot occur until the two ions collide. Because chemical species move faster as the temperature increases, the frequency of collisions increases at higher temperatures, and so the rate of the reaction increases. A termolecular elementary reaction will generally be slow because of the low probability of three reacting species finding each other at the same instant even at high temperatures. Or stated another way, the ΔS for such a process is exceedingly unfavorable. Reactions with a molecularity of four or greater are unknown because there is virtually no chance that four reacting species will find each other at the same time with the correct orientation.

This simple explanation is far from the whole story. Unimolecular reactions also take time even though the reactant does not have to find another reactant. The rate at which two reactants collide in a bimolecular reaction can be calculated based on measured concentrations and temperature. Such calculations show that the rate at which collisions occur is much greater than the rate of the reaction. Clearly not every collision leads to a reaction.

There are several reasons for the disparity between the rate of collisions and the rate of reaction. One reason is that the reaction will not take place unless the two reactants collide in a way that allows the reaction to occur. Consider the example of the carbocation and bromide ion discussed earlier. If the bromide ion does not collide specifically with the positive carbon atom, reaction will not occur.

But the most important reason is related to the energy changes that must take place along the pathway between the reactants and products of a reaction. With the exception of those elementary reactions in which nothing more than formation of a bond occurs, every chemical process involves some bond breaking. As we know, bond breaking requires energy. There is substantial experimental and theoretical evidence that the energy of a reacting system first increases as the reactants are converted to the products. At some point it reaches a maximum energy, which is higher than that of either the reactants or the products, and then it decreases. This point of maximum energy is called the *transition state*. The difference in energy between the reactants and the transition state is called the *energy of activation*, generally abbreviated E_a. This energy pathway is shown in a simplified graphical way in Figure 5.2, which is a plot of how the energy (the *y*-axis) changes as the reaction proceeds (the *x*-axis).

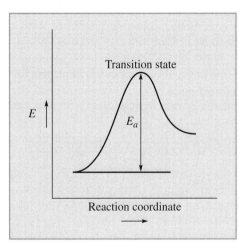

Figure 5.2
An energy diagram

The transition state can be regarded as an energy barrier that the reactants must get over before they can become the products. The height of the barrier is the energy of activation.

The existence of an energy barrier that the reactants must climb to get to products also explains why reactions go faster at higher temperatures. As the temperature increases, the energy of the reactants increases and becomes closer to the energy of the transition state. The barrier is thus smaller. The energy of activation can be measured by measuring the change in reaction rate with a change in temperature.

We often make the analogy between this energy barrier and a mountain peak between two valleys. We must climb to the top and over the peak to get from one valley to the other. The time it takes to climb the peak depends on how high it is above the floor of the starting valley. When the temperature increases, it is as if the altitude of the starting valley had been raised, so that the mountain peak is no longer as high above the floor of the valley. It will take less time to get to the top of the peak.

A very rough rule of thumb that works for many reactions carried out near room temperature is that the rate of the reaction doubles when the temperature is raised by 10 K.

A transition state has an extremely short lifetime and can never be observed directly. But we are still very interested in getting as much information about it as possible because the nature of the transition state provides valuable information about the way in which a reaction takes place.

One way to gain a better understanding of the transition state is to consider it from a thermodynamic point of view. The free energy difference between the reactants and the transition state is called the free energy of activation and is represented by the symbol $\Delta G^{\ddagger}$. As we know (Section 4.4), both ΔH and ΔS contribute to ΔG expressed by the relationship

$$\Delta G = \Delta H - T\Delta S$$

We can define an enthalpy of activation and an entropy of activation, which have the same relationship to $\Delta G^{\ddagger}$.

$$\Delta G^{\ddagger} = \Delta H^{\ddagger} - T\Delta S^{\ddagger}$$

The symbol $\ddagger$ is called a double dagger and is used to designate the transition state or parameters related to the transition state. All three of these quantities, which are called activation parameters, can be measured by measuring the change in the rate of a reaction with temperature.

The relative magnitude of these enthalpy changes and entropy changes provides information about transition states and even reaction mechanisms.

For example, ΔH is related to bond making and bond breaking. The larger the value of $\Delta H^{\ddagger}$, the more bond breaking is occurring in the transition state, since ΔH for bond breaking is positive.

The entropy change ΔS is related to organization. A negative $\Delta S^{\ddagger}$ means that the transition state is more organized than the reactants. A positive $\Delta S^{\ddagger}$ means the transition state is less organized than the products. If two species are coming together in a reaction, we expect a negative $\Delta S^{\ddagger}$ because the transition state will be more organized than the two separate reactants. If one species is coming apart to form two or more products, we expect a positive $\Delta S^{\ddagger}$. In this transition state the reactant is starting to come apart, and the transition state is therefore more disorganized than the reactant.

5.4 | Reaction Coordinate Diagrams

Crude graphs such as the one shown in Figure 5.2 are usually called reaction coordinate diagrams or energy diagrams. They can give useful information about some important characteristics of a reaction in a concise way. As a result, they are used very frequently by organic chemists.

Neither axis of the graph is rigorously defined. The y-axis is labeled energy or E, which means potential energy. But narrower definitions of energy, such as free energy or even enthalpy, are often used when convenient. Because almost all organic reactions are enthalpy controlled under ordinary conditions, we tend to make enthalpy (heat) the y-axis. While the idea of a "reaction coordinate" can be rigorously defined, for our purposes we can think of it simply as the extent to which the reaction occurs. A reaction coordinate diagram is thus a plot of how the energy (or enthalpy) of the chemical system changes as the reaction proceeds.

Reaction coordinate diagrams can be drawn for elementary reactions or for complex reactions that proceed through two or more elementary steps. They can provide qualitative information about the relative energy of reactants and products. When the products are higher on the y-axis than the reactants, as in Figure 5.2, the reaction is endothermic. When the products are lower on the y-axis than the reactants, the reaction is exothermic, as in Figure 5.3. If we have the value of ΔH of the reaction, we can place the reactants and products anywhere on the vertical axis as long as the difference between them is the known ΔH.

Figure 5.3

Reaction coordinate diagram for an exothermic reaction

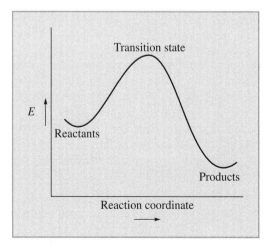

The relative height of the maximum of the curve indicates the energy difference between the reactants and the transition state. This energy difference is the energy of activation, which provides important information about the rate of the reaction.

Reaction coordinate diagrams are also a very convenient way to convey information about a reaction mechanism. Each step in a reaction mechanism is an elementary reaction, which has reactants, products, and a transition state. The products of the first elementary reaction are the reactants for the next one in the reaction mechanism sequence. We can draw a continuous curve to describe the progress of the reaction along the reaction coordinate.

For example, Figure 5.4 shows a reaction coordinate diagram for a reaction that proceeds by a two-step mechanism. Since each step has a transition state, the diagram has two maxima.

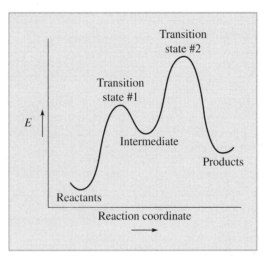

The diagram conveys a great deal of information about the reaction beyond just that it proceeds by a two-step mechanism. The reaction is endothermic. There is a relatively high-energy intermediate that forms from the reactants and is then converted to the products. Transition state #2 is the highest-energy point on the curve, which means that step #2 is the rate-determining step.

Example 5.3

Draw a reaction coordinate diagram for a two-step exothermic reaction in which the first step is rate determining.

Solution

The products will be lower in energy than the reactants in an exothermic reaction. The transition state for the first step will be higher in energy than that for the second step. The diagram is shown in Figure 5.5.

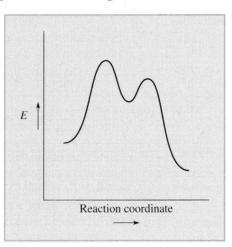

Figure 5.5
Reaction coordinate diagram for a two-step exothermic reaction

Because these diagrams combine thermodynamic and kinetic information in one compact picture, they can be very helpful in understanding some of the principles that govern reactions in complicated systems.

5.4A Reactivity and Selectivity

It is not uncommon to encounter situations in organic chemistry where a given set of reactants can form different isomeric products. In studying such a reaction system, one of the most important things we would like to know is what determines the relative amounts of the products.

For example, reagents such as Cl_2 and Br_2 react with many organic compounds to form products in which an H of the organic compound is replaced by Cl or Br. The simplest example of such a reaction is

$$CH_4 + Cl_2 \rightarrow CH_3Cl + HCl$$

Compounds that are more complicated than methane, however, will generally be able to form more than one isomeric product. For example, propane, C_3H_8, can react with Cl_2 to form two isomeric chloride products, as shown in Figure 5.6.

Figure 5.6
Monochlorination products of propane

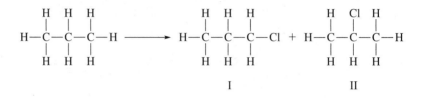

The same reaction can take place with Br_2 to form the corresponding isomeric bromides.

Being able to predict the relative amounts of each isomeric product is of great practical interest. Not surprisingly, the product distribution depends on the conditions under which the reaction is carried out. But what is consistently observed is that under a given set of reaction conditions, the isomer with the halogen on the middle carbon (II) is always found in greater relative amount with Br_2 than with Cl_2.

To explain this observation, we need to answer a number of questions:

1. Is the formation of Product II faster than the formation of Product I? The answer is yes.
2. Why? The simple answer without going into all the details of the mechanism is that the intermediate that leads to the formation of Product II is lower in energy than the intermediate that leads to the formation of Product I.
3. How does this intermediate form? A halogen atom reacts with the propane.
4. Under a given set of reaction conditions, which halogen reacts faster? The Cl atom reacts faster than the Br atom. Stated another way a Cl atom is more reactive than a Br atom.

We know from our experience that when something is more reactive, it is less selective. For example, when you are very hungry (very reactive), you will eat almost anything (less selective). When you are not very hungry, you are much more selective about what you will eat. The same principle applies to chemical reactions, and we can show why, using reaction coordinate diagrams.

The relative energy of the transition state determines the relative rate of a reaction. Unfortunately, we cannot observe a transition state directly or make measurements on it. We can only infer what its structure might be. The transition state must have some structural features of the reactants and some structural features of the products. The logical assumption is that the transition state most resembles the things to which it is closest in energy. In an endothermic reaction, such as the one shown in Figure 5.2, the transition state is closer in energy to the products than to the reactants and will resemble the products more than the reactants. In an exothermic reaction, such as the one shown in Figure 5.3, the transition state is closer in energy to the reactants and will resemble the reactants more than the products. This assumption is consistent with the behavior of reacting chemical systems.

Figures 5.7 and 5.8 show reaction coordinate diagrams for a chemical system in which one set of reactants can form two different products, one more stable than the other.

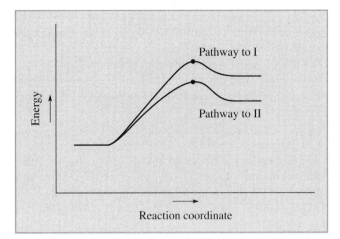

Figure 5.7
Endothermic reaction that can form two products

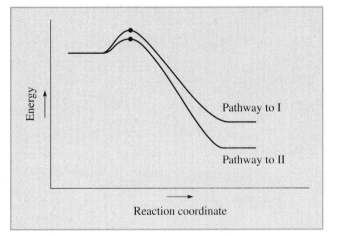

Figure 5.8
Exothermic reaction that can form two products

The reaction in Figure 5.7 is endothermic, and the transition states will therefore resemble the products. The difference in the energy of the products is substantially reflected in the difference in energy between the transition states that leads to each of them. The greater the difference in the energy of these transition states, the greater the difference in the rates of their formation and the greater the fraction of the more stable product that forms.

The reaction in Figure 5.8 is exothermic, and the transition states will therefore resemble the reactants. Since both products are formed from the same reactants, these transition states will be closer in energy to each other than in the endothermic case. Therefore the relative amounts of the products will be closer to each other.

To summarize, the more stable product is formed in greater relative amount in an endothermic elementary reaction than in a corresponding exothermic elementary reaction. The exothermic reaction represents a situation of higher reactivity; it is faster because it has a lower transition state, one closer in energy to the reactants.

The reaction of a chlorine atom with propane is exothermic. The reaction of a bromine atom with propane is endothermic. Bromine is more selective and always gives a great relative amount of the more stable intermediate that leads to product (II) than does chlorine.

Notice, however, that the major product is not necessarily the one from the more stable intermediate. The reaction is one in which a halogen atom replaces a hydrogen. In propane, the replacement of any one of 6 hydrogens gives product I. But there are only 2 hydrogens whose replacement gives product II. So unless the conditions are such that the halogen is very selective, product II may not be the major product.

5.4B Kinetic and Thermodynamic Control

It is not uncommon to encounter situations in organic chemistry where a given set of reactants can form different isomeric products. But when the reaction conditions allow the system to reach equilibrium, we can choose to form either product by adjusting the reaction conditions.

This situation can be represented by the general scheme:

$$A + B \rightleftharpoons C$$

$$A + B \rightleftharpoons D$$

In studying such a reaction system, one of the most important things we would like to know is which product is obtained at a given set of conditions. Two factors seem relevant in determining whether the product is C or D. One factor is the relative stabilities of C and D. The other factor is the relative rates at which C and D form. If the more stable product is also the one that forms fastest, then, unless there are some other factors in play, it will be the major product.

Often, however, the most stable product is not the one that forms fastest. Let us assume that product C forms faster and product D is more stable, as represented in the *modified* reaction coordinate diagram shown in Figure 5.9, and that the formation of each product is reversible.

Figure 5.9

Reaction coordinate diagram for a reaction in which one product, C, forms faster and another product, D, is more stable

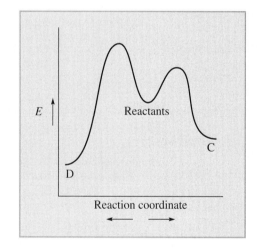

In this situation it is possible to form selectively either C or D by controlling the way in which the reaction is carried out. The product that forms faster, C, is called the product of kinetic control. We can obtain it by allowing the reaction to proceed for only a relatively short time. The more stable product, D, is called the product of thermodynamic control. It is obtained if we allow the system to proceed to equilibrium. Since D is more stable, it will predominate at equilibrium.

The reaction coordinate diagram shown in Figure 5.9 also helps us to understand the buildup of product D. If we examine the energy barriers and thereby the relative rates of the reverse reactions, we can see that while C forms faster, it also decomposes faster when conditions permit the reaction to reverse. When D finally forms, it decomposes to the reactants more slowly than C. Thus, when D forms, it tends to remain.

In summary, reaction coordinate diagrams are a very useful, compact way to describe the energy changes that take place during the course of a chemical reaction. They convey information about the relative energy of reactants and products, transition states, and intermediates. They are used very frequently in organic chemistry.

5.1 The reaction of CH_3Cl with I^- to form $CH_3I + Cl^-$ follows a second-order rate law, first order in each reactant. When a solution is prepared at a given temperature in which the concentration of each reactant is 0.40 M, the rate at which the CH_3Cl is converted to CH_3I is 0.020 M/min. Find the rate of conversion when the concentration of CH_3Cl is 0.60 M and $[I^-]$ is 0.20 M at the same temperature.

5.2 When $H_2(g)$ and $Br_2(g)$, each at a pressure of 0.10 atm, react, the rate at which the $H_2(g)$ disappears is 0.030 atm/min. Find the rate at which $H_2(g)$ disappears when the pressure of each gas is 0.08 atm. Use the rate law for this reaction given in Section 5.1.

5.3 Use the data in the following table to find the rate law for the reaction of C_2H_4O (acetaldehyde) with chlorine:

H—C—C—H + Cl₂ ⟶ H—C—C—H + HCl

[C_2H_4O] M	[Cl_2] M	Rate of C_2H_3ClO formation (M/sec)
0.30	0.30	0.060
0.20	0.30	0.040
0.10	0.20	0.020
0.20	0.10	0.040

5.4 Each of the intermediates in Figure 5.1 can be converted to ethane, C_2H_6, by a reagent that provides the species that needs to be added. For each of the intermediates identify the species needed and suggest a reagent that might provide it.

5.5 The reaction of C_2H_2, acetylene, with HBr in the gas phase to form vinyl

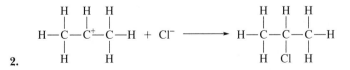

bromide, C_2H_3Br,
order rate law, first order in acetylene and second order in HBr. Propose a mechanism.

5.6 The reaction $C_3H_6 + HCl \rightarrow C_3H_7Cl$ may take place by a two-step mechanism under certain conditions. In the first step H^+ adds to the C_3H_6 to form a cation:

1.

In the second step the cation reacts with Cl^-.

2.

Write the rate law corresponding to this mechanism if **(a)** the first step is rate determining or **(b)** the second step is rate determining. Which is more likely?

5.7 When formaldehyde, $CH_2{=}O$, is dissolved in water, it very rapidly adds H_2O to form CH_4O_2. In solutions that are acidic, the reaction follows a second-order rate law, first order in $CH_2{=}O$ and first order in H^+. Suggest a reaction mechanism in which each step is an acid–base reaction and identify the acid and base in each step.

5.8 Both the ethyl radical, $C_2H_5\cdot$, and the propyl radical, $C_3H_7\cdot$, react with Cl_2 in the gas phase to form $Cl\cdot$ and propane. Predict which reaction will be faster under the same conditions and give two reasons to support your answer.

5.9 Consider the rate-determining step of the reaction of ethyl alcohol and isopropyl alcohol with HBr described in Section 5.2. Which one has the most unfavorable $\Delta H^{\ddagger}$, and which one has the most unfavorable $\Delta S^{\ddagger}$. Explain.

5.10 Draw reaction coordinate diagrams for the following reactions:
    (a) A two-step exothermic reaction in which the second step is rate determining
    (b) A two-step endothermic reaction involving a high-energy intermediate, in which the first step is rate determining
    (c) A three-step exothermic reaction in which the first step is rate determining

5.11 Draw reaction coordinate diagrams for the reactions of ethyl alcohol and isopropyl alcohol with HBr described in Section 5.2. [Hint: Use the bond energy values in Tables 1.3 and 4.1 as necessary.]

Appendix

Solutions to End of Chapter Exercises

1.1 **(b), (d), (f), (h)**

1.2 **(b)** N, **(d)** Cl, **(f)** P, **(h)** Al

1.3 For elements 3 to 6, preferred number of bonds = number of valence electrons; for elements 7 to 9, preferred number of bonds = 8 − number of valence electrons.

1.4 The preferred number of bonds = 6 − number of 2p electrons

1.5

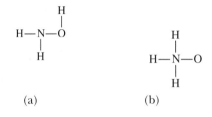

(a)   (b)

(a) Corresponds to the more stable molecule

1.6

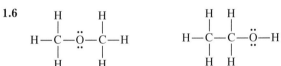

dimethyl ether   ethyl alcohol

1.7

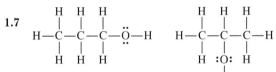

propyl   isopropyl

1.8

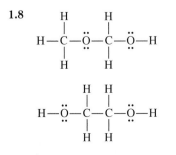

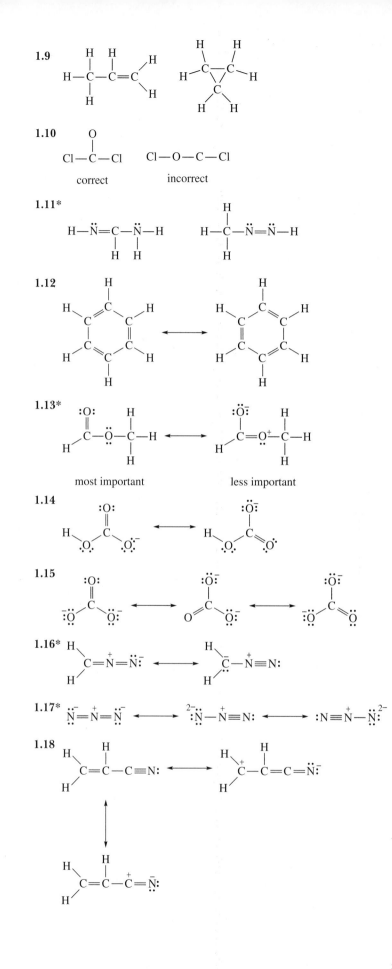

1.19*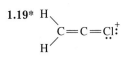

1.20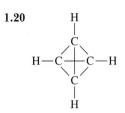

1.21 (a)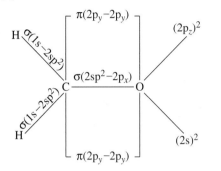

(b)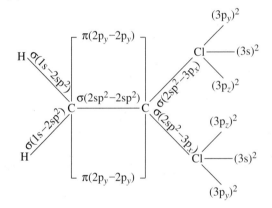

(c)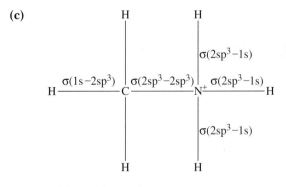

All C—H the same

(d)

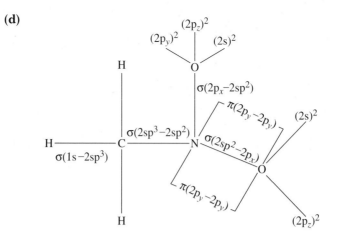

All C—H the same

(e)

All C—H the same

1.22

(a)

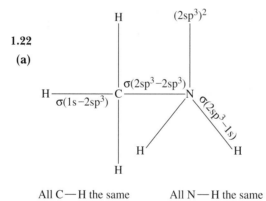

All C—H the same   All N—H the same

(b)

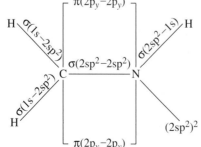

(c)

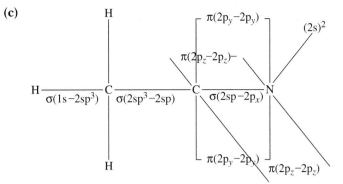

All C—H the same

1.23

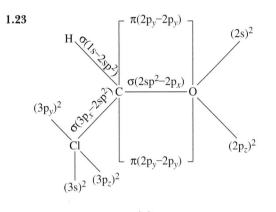

(a)

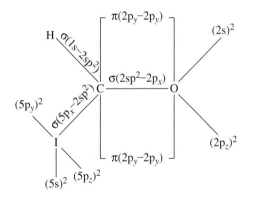

(b)

1.24

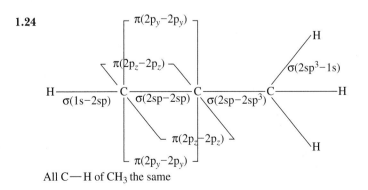

All C—H of CH$_3$ the same

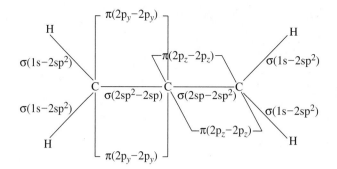

1.25

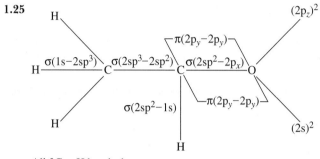

All 3C—H bonds the same

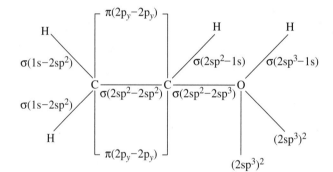

1.26 (hint) **(a)**   **(b)**

1.27 **(a)** Tetrahedral **(b)** tetrahedral **(c)** linear **(d)** tetrahedral **(e)** trigonal planar **(f)** trigonal planar

1.28 **(b)** **(a)** **(d)** **(c)** **(e)**

1.29* (hint)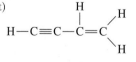

1.30 (hint) **b, c, d**

1.31* (hint)

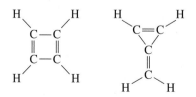

1.32 (a)

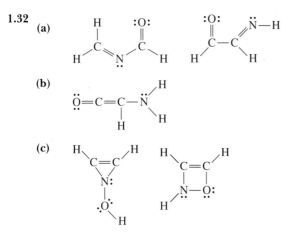

(b)

(c)

1.33 The electron pair in each of the N—F bonds is relatively far from the N because the F is so electronegative. The FNF bond angles can close down to minimize repulsions with the lone pair on N (VSEPR). The electronegative F will use a hybrid orbital with more p character. The increased p character results in smaller bond angles (hybridization).

1.34 C_2H_2, C_2H_4, C_2H_6. The C_2H_2 has the shortest C—H bond, and the C_2H_6 has the longest C—H bond.

1.35 CH_3F, CH_3Cl, CH_4S, CH_5P

1.36 C—P, C—S, C—N, C—O

1.37 Since a C—Cl bond is stronger than a C—N bond and a C—Br bond is stronger than a C—P bond, it appears that electronegativity differences are more important than size differences.

1.38 CH_3SH, CH_3Br, CH_3Cl, CH_3F

1.39*

2.1 8(equal to the number of σ bonds). 5. There are two kinds of C—C σ bonds and three kinds of C—H bonds (three bonds of one kind, two bonds of another, and one bond of the third)

2.2 9(equal to the number of σ bonds). 2. All the C—C bonds are the same and all the C—H bonds are the same.

2.3 These orbitals differ from those of ethylene because there is more electron density near the O. The drawing should reflect that.

2.4

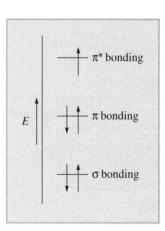

The instability results from an expanded octet on one of the carbons.

2.5 The pair of electrons on the hydride ion is in the 1s orbital. This pair is donated to the vacant 2p orbital on the $2sp^2$-hybridized C of the methyl cation. The product of the reaction is methane, CH_4, so the electron pair is in a $2sp^3$ orbital.

2.6 The pair of electrons on the hydride ion is in the 1s orbital. This pair is donated to the π* antibonding orbital of C_2H_4. The product of the reaction is an anion with the formula $CH_3CH_2 \ominus$. The electron pair is in a $2sp^3$ hybrid orbital.

2.7.* Electron density between the nuclei has favorable Coulombic attractions between the negative electrons and the positive nuclei. They also shield the nuclei from each other, lowering the unfavorable Coulombic repulsion between the two positively charged nuclei.

2.8 When two waves combine constructively, they make a bigger wave with higher crests. The amplitude of the wave is a measure of the probability of the electron being in that region of space.

2.9 A three-atom system with an unhybridized 2p orbital at every atom has three molecular orbitals spread over the three atoms. One is the bonding orbital, which is of lower energy than the 2p orbitals, as well as an ordinary π bonding orbital. The second is the nonbonding orbital, which has the same energy as the 2p orbitals, and the third is the higher-energy antibonding orbital. This system has only three electrons that need to be accommodated by these orbitals. A pair of electrons fills the bonding orbital, which results in a decrease in their energy. The remaining electron is in the nonbonding orbital, which results in no energy change. The net result is a lower energy.

2.10 The molecular orbitals are those described in the solution to Exercise 2.9. The anion has four electrons that need to be accommodated by these orbitals. One pair is in the bonding orbital; the other pair is in the nonbonding orbital. The net result is a lower energy.

2.11 The central C is a 2sp hybrid. It has two unhybridized 2p orbitals that are mutually perpendicular. One of these orbitals is parallel to the 2p orbital on one of the terminal C atoms, and the other is parallel to the 2p orbital on the other terminal C atom. Since these orbitals on the central carbon are perpendicular, the 2p orbital on each of the terminal carbon atoms must also be perpendicular to each other, as shown in this figure.

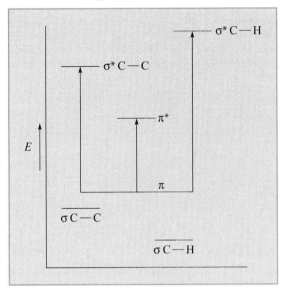

In order to have conjugation, all three of the 2p orbitals must be parallel.

2.12 The $\pi \rightarrow \pi^*$ is at the lowest energy and therefore at the longest wavelength. There are two types of σ bonds, the C—H and the C—C. The C—H bond is stronger, so its σ bonding orbital is at lower energy than that of the C—C bond. Therefore the σ* orbital

of the C—H must be at higher energy than that of the C—C. Thus there are two $\pi \rightarrow \sigma^*$ transitions. The one in which the π electron jumps to the σ* of the C—H is of higher energy and therefore at the shortest wavelength. These transitions are shown in the diagram.

2.13 Since there are two π bonding orbitals and two π* antibonding orbitals, there are four possible $\pi \rightarrow \pi^*$ transitions. The one of greatest energy is the $\pi_1 \rightarrow \pi_4{}^*$, and the one of lowest energy and therefore of longest wavelength is the $\pi_2 \rightarrow \pi_3{}^*$. The other two are the $\pi_1 \rightarrow \pi_3{}^*$ and the $\pi_2 \rightarrow \pi_4{}^*$. There will be two $\pi \rightarrow \sigma^*$ transitions for the C—H σ*. In one it is the electron from the π_2 orbital that jumps and in the other from the π_1 to the σ*. This latter transition is the one with the greatest ΔE, since the π electrons in π_1 are the lowest-energy π electrons and the C—H bond is the strongest bond, with the lowest σ bonding orbital and therefore the highest-energy σ* antibonding orbital. So this one is the shortest-wavelength transition. There are also two types of C—C σ bonds, one type a single bond between the two interior carbons and the other type the single bond part of the double bonds. Since there are two types of π electrons and two different σ* orbitals, there will be four more $\pi \rightarrow \sigma^*$ transitions for the C—C σ*. These transitions are shown in the diagram:

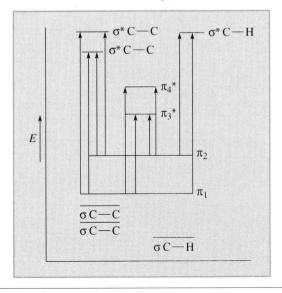

2.14* There are three types of σ bonds in this molecule, the C—H, the N—H, and the σ component of the double bond. Since the C—H bond is the strongest of these three bonds, it has the lowest-energy σ bonding orbital and therefore the highest-energy σ* antibonding orbital. The π electrons are of lower energy than the lone pair because they are in a bonding orbital. So the transition of greatest energy and shortest wavelength is the $\pi \rightarrow \sigma^*$ transition to the σ* of the C—H. There will also be $\pi \rightarrow \sigma^*$ transitions to the N—H σ* and the C—N σ*. There will be three $n \rightarrow \sigma^*$ electronic transitions in which one of the lone-pair electrons jumps to one of the three different σ* orbitals. In addition there is an $n \rightarrow \pi^*$ transition, which is of lower energy than any of the transitions to the σ* orbitals. It is the transition of longest wavelength. These transitions are shown in the diagram:

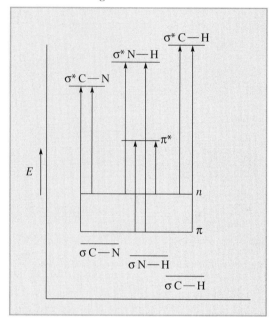

2.15 9.1×10^{-19} J/mol or 546 kJ/mol, using $\Delta E = hc/\lambda$ and $N = 6.0 \times 10^{23}$ mol^{-1}.

2.16 The ΔE for this $\pi \rightarrow \pi^*$ transition is 1.1×10^{-18} J, using $\Delta E = hc/\lambda$.

The strength of the π bond is the ΔE between the π bonding orbital and the nonbonding p orbital. This value is half the ΔE between the π and π* orbitals. Thus the bond strength is 5.2×10^{-19} J or 312 kJ/mol, using $N = 6.0 \times 10^{23}$ mol^{-1}.

2.17

3.1 (a) Both: PO_4^{3-}, $H_2PO_4^-$; (b) base

$$Cl-\overset{\overset{\displaystyle H}{|}}{\underset{\underset{\displaystyle Cl}{|}}{N^+}}-Cl$$

; (c) base $HO-C^+=O$;
(d) acid Cl_3C^-; (e) neither; (f) base H_2C^+-OH, while an acid in theory because it has an H, it is so weak that it need not be considered as an acid; (g) both NH_4^+, NH_2^-; (h) both $CH_3OH_2^+$, CH_3O^-; (i) neither.

3.2

$$\overset{\displaystyle H\diagdown \quad +\diagup Cl}{\underset{\underset{\displaystyle H}{|}}{O}}$$

3.3 (a) $HCO_3^- + HI \rightleftharpoons H_2CO_3 + I^-$

(b) $HF + CN^- \rightleftharpoons HCN + F^-$

(c) $HCl + PH_3 \rightleftharpoons PH_4^+ + Cl^-$

(d) No reaction

(e) $CH_3CO_2H + CH_3NH_2 \rightleftharpoons CH_3CO_2^- + CH_3NH_3^+$

(f) $CH_3O^- + CH_3NH_3^+ \rightleftharpoons CH_3OH + CH_3NH_2$

(g) No reaction

(h) $CH_2OH^+ + NH_3 \rightleftharpoons CH_2=O + NH_4^+$

3.4 (a) HCO_3^-/H_2CO_3 and HI/I^-

(b) HCN/CN^- and F^-/HF

(c) HCl/Cl^- and PH_3/PH_4^+

(e) $CH_3CO_2H/CH_3CO_2^-$ and $CH_3NH_2/CH_3NH_3^+$

(f) CH_3O^-/CH_3OH and $CH_3NH_3^+/CH_3NH_2$

(h) $CH_2OH^+/CH_2=O + NH_3/NH_4^+$

3.5

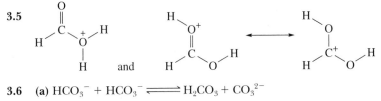

3.6 (a) $HCO_3^- + HCO_3^- \rightleftharpoons H_2CO_3 + CO_3^{2-}$

      base   acid   conj acid   conj base

(b) $NH_3 + NH_3 \rightleftharpoons NH_4^+ + NH_2^-$

      base   acid   conj acid   conj base

(c) No   (d) No

(e) $HSO_4^- + HSO_4^- \rightleftharpoons H_2SO_4 + SO_4^{2-}$

      base   acid   conj acid   conj base

(f) No

3.7

$$H_2N-\overset{\overset{\displaystyle H}{|}}{\underset{\underset{\displaystyle H}{|}}{C}}-\overset{\overset{\displaystyle H}{|}}{\underset{\underset{\displaystyle H}{|}}{C}}-\overset{+}{N}H_3$$

3.8 $HSO_4^- + PO_4^{3-} \rightleftharpoons SO_4^{2-} + HPO_4^{2-}$

3.9 $HNO_3 + OH^- \rightleftharpoons H_2O + NO_3^-$ largest

$H_2O + NO_3^- \rightleftharpoons HNO_3 + OH^-$ smallest

3.10 (a) Largest K;   (b) smallest K

3.11 (a) p$K = 1$;   (b) p$K = -1.91$;   (c) p$K = -4.62$;

    (d) p$K = 6.08$;   (e) p$K = 0.8$ (f) p$K = 15$

3.12 (a) H_2Te, H_2Se, H_2S, H_2O;   (b) HCl, H_2S, PH_3, SiH_4;

    (c) $HClO_4$, H_2SO_4, H_3PO_4, H_3PO_3, H_3PO_2 (d) z, v, y, w, x

3.13 (a) SbH_3, AsH_3, PH_3, NH_3 (b) w, y, z, v, x

    (c) ClO_2^-, BrO_2^-, IO_2^- (d) BrO_4^-, BrO_3^-, BrO_2^-, BrO^-

3.14 CH_3^+, N_2O, NO^+, Cl^+, $Al(OH)_3$

3.15 Cl^+, PH_4^+, BeF_2

3.16 C_2H_4, CH_3N, CH_2O

3.17 HBr (acid) + C_2H_4 (base) $\rightleftharpoons$ $C_2H_5^+$ + Br^-

    $C_2H_5^+$ (Lewis acid) + Br^- (Lewis base) $\rightleftharpoons$ C_2H_5Br

3.18 H^+ + $CH_2{=}O$ $\rightleftharpoons$ $H_2C{=}OH^+$ $\longleftarrow$ $H_2C^+{-}OH$

3.19 The OH^- is more nucleophilic because it is more basic.

3.20 Since fluoride ion is much smaller than iodide ion, it is much more strongly solvated by hydrogen bonding to water. This strong solvation reduces its nucleophilicity. If the solvent does not have OH or NH bonds, it cannot hydrogen bond and therefore does not solvate the anions very well.

Chapter 4

4.1 **(a)** and **(b)** since heat must be absorbed to melt ice; freezing of water, the reverse process, must give off heat; **(c)** a type of combustion reaction; **(f)** we give off heat when we exercise. Not exothermic are **(d)** since vaporization requires the absorption of heat and **(e)** which is the reverse of combustion.

4.2 **(a)** A bond is breaking, no bonds are being made; **(b)** the reverse of hydrogenation, which is exothermic; **(d)** the reverse of combustion, which is exothermic; **(f)** this reaction does not take place despite the favorable entropy. Not endothermic are **(c)** the reaction of a strong acid with a strong base and **(e)** which takes place despite the unfavorable entropy.

4.3 **(a)** It is C_3H_8 because one mole of C_3H_8 has a greater mass than one mole of C_2H_6.

(b) It is $CH_2\!=\!CHCH_2CH\!=\!O$ because it is unconjugated and therefore less stable than $CH_3CH\!=\!CH\!-\!CH\!=\!O$, which is its isomer and therefore has the same molecular weight.

4.4 275 kJ/mol

4.5* Br—F in BrF_3 = 197 kJ/mol, in BrF_5 = 184 kJ/mol. The bond is weaker in BrF_5 because of the destabilization resulting from crowding 5 atoms around a single central one.

4.6 $F_2 = -551$ kJ/mol; $Cl_2 = -144$ kJ/mol; $Br_2 = -108$ kJ/mol; $I_2 = -16$ kJ/mol.

The most important factor in determining the favorability of reaction is the strength of the C—X bond. The stronger it is, the more favorable the reaction.

4.7 **1.** $\Delta H = +23$ kJ/mol; **2.** $\Delta H = +63$ kJ/mol $- 40$ kJ/mol $= +23$ kJ/mol. As expected the value of ΔH is independent of the path.

4.8 499 kJ/mol

4.9 36 kJ/mol

4.10* 82 kJ/mol

4.11 -182 kJ/mol. Resonance is more important in acrolein because of the contributing structure with the negative charge on the very electronegative oxygen.

4.12 Cyclopropane is less stable than propene because its structure requires substantial distortion of the C—C—C bond angles from 109.5° to 60°. The destabilization is called angle strain. It destabilizes the cyclopropane by 115 kJ/mol.

4.13* -188 kJ/mol

4.14* -112 kJ/mol

4.15 4849 kJ/mol. Each compound in the series differs by CH_2 from the previous one and by 657 kJ/mol in its heat of combustion. Heptane has one more CH_2 than hexane, and so the prediction is that its heat of combustion increases by 657 kJ/mol.

4.16 **(a)** Negative, four molecules to two; **(b)** negative, major effect is converting one molecule of a gas to a liquid; **(c)** positive, major effect is converting one molecule of a liquid to a gas; **(d)** positive, diatomic molecules with the same two nuclei are more organized than those with different nuclei, likely to be a small effect; **(e)** positive, two molecules from one, even though one is a solid.

4.17 Cyclopropane, because of its ring structure, is a relatively rigid molecule with virtually no freedom of movement around its bonds. Opening the ring allows the system to have more flexibility of motion and become more disorganized. The ring-opening reaction is the one with the greater ΔS.

4.18 **(a)** Bonds broken: C—Br 285, N—H 391 = 676 kJ/mol

Bonds made: C—N 305, H—Br 366 = 671 kJ/mol

$\Delta H = 676 - 671 = 5$ kJ/mol

There are two reactant molecules and two product molecules. ΔS should be zero or very close to zero.

The reaction is not likely to be spontaneous because it is slightly endothermic with no significant entropy change.

(b) This reaction is the combustion of acetylene, a very exothermic reaction used in oxyacetylene torches. The small unfavorable ΔS will not have much effect on its spontaneous nature.

(c) Bonds broken: C—H 413, I—I 151 = 564 kJ/mol

Bonds made: C—I 218, H—I 299 = 517 kJ/mol

$\Delta H = 517 - 564 = 47$ kJ/mol

There are two reactant molecules and two product molecules. ΔS should be zero or very close to zero.

The reaction is not spontaneous because it is endothermic with no significant entropy change.

(d) Bonds broken: C=C 615, F—F 158 = 773 kJ/mol

Bonds made: 2 C—F 489 × 2 + C—C 346 = 1324 kJ/mol

$\Delta H = 773 - 1324 = -551$ kJ/mol

The reaction is spontaneous, notwithstanding the small unfavorable entropy change, because it is very exothermic.

(e) Bonds broken: 2 C—Cl 328 × 2 + C=C 615 = 1271 kJ/mol

Bonds made: C—Cl 243, C≡C 812 = 1055 kJ/mol

$\Delta H = 1271 - 1055 = 216$ kJ/mol

The reaction is not spontaneous because it is endothermic with a relatively small entropy change. Higher temperature would be required to carry out the reaction.

4.19 This reaction is enthalpy controlled because the ΔS for each step is relatively small.

(a) The first step is very unfavorable because it is highly endothermic. A C—O is broken and no bond forms, $\Delta H = 358$ kJ/mol. Although the second and third steps are both exothermic ($\Delta H = 299 - 463 = -164$ kJ/mol for step 2 and $\Delta H = -218$ kJ/mol), the first step is so unfavorable that this mechanism is unlikely to be correct.

(b) The first step is also unfavorable because it is endothermic ($\Delta H = (358 + 299) - (463) = 194$ kJ/mol) although less so than the first step in (a). Although ΔS is positive for this step and the next step is very exothermic, those factors do not seem great enough to overcome the still significant endothermicity.

(c) The ΔS for this mechanism is quite unfavorable (negative) because it requires bringing two things together with just the right orientation for everything to switch simultaneously:

$$CH_3—OH$$
$$I\underline{\hspace{2cm}}H \;\; .$$

But the reaction is exothermic: $\Delta H = (358 + 299) - (218 + 463) = -24$ kJ/mol. So while this mechanism still has some difficulties and may not be correct, it appears to be the best of the three choices.

5.1   0.015 M/min

5.2   0.021 atm/min

5.3   Rate = $k[C_2H_4O]$

5.4   (a) The ethyl cation needs a hydride ion, H^-, which can be provided by a metal hydride such as NaH.
(b) The ethyl anion needs a proton, which can be provided by any Brønsted acid.
(c) The ethyl radical needs an H atom, which can be provided by compounds that contain H. Simplest ones are binary hydrides of the nonmetals, such as HX.

5.5   Simplest is a termolecular elementary reaction in which one HBr points its H end at one of the carbons and the other HBr points its Br end at the other carbon.

$$
\begin{array}{c}
H\!-\!Br \\
| \\
H\!-\!C\!\equiv\!C\!-\!H \\
| \\
Br\!-\!H
\end{array}
$$

5.6   **(a)** Rate = $k[H^+][C_3H_6]$; **(b)** Rate = $k[H^+][C_3H_6][Cl^-]$

(a) is more likely to be slower because a bond is being made and a bond is being broken.
(b) is likely to be faster because a bond is being made, but no bond is being broken.

5.7   **1.** $CH_2{=}O$ (Brønsted base) + H_3O^+ (Brønsted acid) $\rightleftharpoons$ $CH_2{=}OH^+$ (conj acid) + H_2O (conj base)

The conjugate acid of $CH_2{=}O$ has two important contributing structures:

$CH_2{=}OH^+ \rightleftharpoons {}^+CH_2{-}OH$

It is easier to visualize the next step with the second contributing structure

2. ${}^+CH_2{-}OH$ (Lewis acid) + H_2O (Lewis base) $\rightleftharpoons$

3. $H_2O^+{-}CH_2{-}OH$ (Brønsted acid) + H_2O (Brønsted base) $\rightleftharpoons$

$$
\begin{array}{c}
OH \\
| \\
H\!-\!C\!-\!OH \\
| \\
H
\end{array}
$$

(conj base) + H_3O^+ (conj acid)

Since H_2O is present in large excess, either the first or the second step could be rate determining

5.8   The reaction with the ethyl radical should be faster. Since the ethyl is lighter, it will move faster, resulting in more frequent collisions. The likelihood of the Cl_2 colliding with the electron deficient C is greater for ethyl because it has fewer carbons.

5.9   The rate-determining step for the ethyl alcohol is bimolecular. The transition state will require the two separate reactants to come together. This increase in organization results in an unfavorable $\Delta S^{\ddagger}$. In the transition state the C$-$O is partly broken and the C$-$Br bond is partly formed indicating that $\Delta H^{\ddagger}$ will not be especially unfavorable.

The rate-determining step for the isopropyl alcohol is unimolecular; the C$-$O bond is beginning to break. Thus the reactant is on the way to forming two products. This increase in disorganization results in a relatively favorable $\Delta S^{\ddagger}$. Because a bond is partially broken in this step and no new bond is forming, the $\Delta H^{\ddagger}$ will be especially unfavorable.

5.10

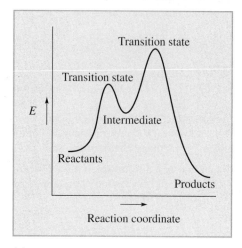

(a)

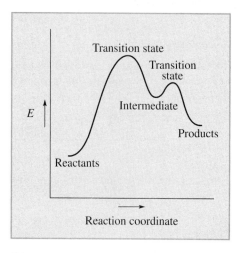

(b)

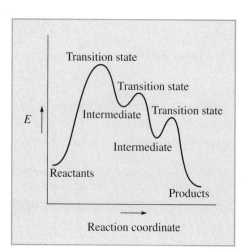

(c)

5.11 Based on average bond energies, each reaction has $\Delta H = -24$ kJ/mol.

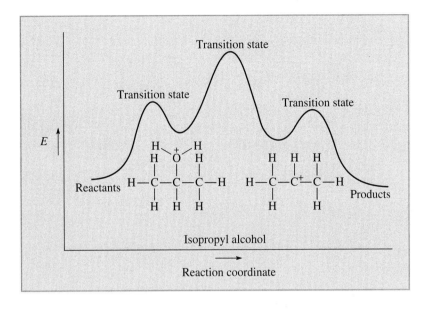

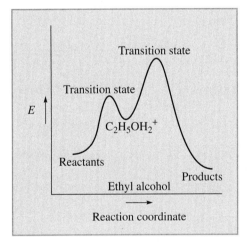